Agricultural Drawing and the Design of Farm Buildings

by Thomas E. French and Frederick E. Ives

with an introduction by Jackson Chambers

This work contains material that was originally published in 1915.

This publication is within the Public Domain.

This edition is reprinted for educational purposes
and in accordance with all applicable Federal Laws.

Introduction Copyright 2018 by Jackson Chambers

IMPORTANT NOTE & DISCLAIMER

IMPORTANT NOTE :

As with all reprinted books of this age that are intended to perfectly reproduce the original edition, considerable pains and effort had to be undertaken to correct fading and sometimes outright damage to existing proofs of this title.

At times, this task can be quite monumental, requiring an almost total rebuilding of some pages from digital proofs of multiple copies. Despite this, imperfections still sometimes exist in the final proof and may detract slightly from the visual appearance of the text.

Some images may suffer from reduced quality due to anomalies in the original scan.

DISCLAIMER :

Due to the age of this book, some methods or practices may have been deemed unsafe or unacceptable in the interim years. In utilizing the information herein, you do so at your own risk.

We republish antiquarian books with no judgment or revisionism, solely for their historical and cultural importance, and for educational purposes.

Self Reliance Books

Get more historic titles on animal and stock breeding, gardening and old fashioned skills by visiting us at:

introduction

Here at **Self-Reliance Books** we are dedicated to bringing you the best in *dusty-old-book-knowledge* – this time, an extremely useful old book filled with the techniques and info you need to make your own plans for building structures.

This special edition of **Agricultural Drawing and the Design of Farm Buildings** was written by Thomas E. French and Frederick E. Ives, and first published in 1915, making it just over a century old.

The book contains sections on *Theory and Technique, Working Drawings, Farm Structures, Maps and Topographical Drawing, Pictorial Drawing, Construction Data*, and more.

An absolutely essential book for anybody, especially those in Sustainable Agriculture, Homesteading, and Off-the-Grid Living who want to learn the techniques and skills of drafting and drawing plans for their own utility buildings.

~ Jackson Chambers

State of Jefferson, April 2018

PREFACE

As the title implies, this book is intended primarily for students in agriculture and agricultural engineering. Engineering Drawing is a required subject in practically all college agricultural courses. It is not given in these courses with the idea of making professional draftsmen, but is regarded as an important subject for increasing the efficiency of the farm owner or manager, by giving him what is in reality a new language in which to express and record his ideas.

Aside from mechanics and builders, there is no class to whom the value of technical drawing should appeal with as much force as to the progressive farmer. His literature is full of illustrations and technical sketches, which to be read intelligently require a knowledge of technical drawing. Government bulletins, State bulletins, agricultural periodicals and books, even trade and machinery catalogues, cannot be fully understood without this knowledge. In order to build properly, or to pass upon a set of plans, he should be able to read architectural drawings. The man with the ability to draw "to scale" can plan his buildings, "take off" his bill of materials, estimate the costs, and mentally see the finished structure before it is built. He can make sketches of broken parts of machinery, or of special pieces which he wishes to have made, he can make a layout of his buildings or a plat of his farm. In short, he has an asset of distinct advantage and value.

This book is a text book* rather than a "course in drawing." The principles and processes involved are described and illustrated, and a variety of problems of various kinds and of progressive difficulty have been arranged, with outlines for a considerable number of additional ones, not only giving suggestions to the farm owner, but also supplying class material, which may thus be varied from year to year. Drawing courses vary in length, and the instructor may make his choice from these numerous problems to cover the different divisions of the subject matter included in the text.

These problems have been selected for their practical value, and all are dependable in design. Many are from work designed and built by the authors.

The freehand method of introducing projection drawing has been used with marked success in agricultural classes.

A number of formulas, tables, etc., have been grouped in one chapter, to give in convenient form information necessary in designing some structures; and other items of miscellaneous information useful in drawing and designing have also been included. In the last chapter is given a list of books and bulletins on allied subjects.

The assistance of Mr. C. L. Svensen and Mr. W. D. Turnbull is gratefully acknowledged.

The authors will be glad to cooperate with teachers using the book as a text book, and to suggest or furnish supplementary problems.

*Some of the material in it has been condensed from the larger text book "A Manual of Engineering Drawing."

COLUMBUS,
August 10, 1915.

T. E. F.
F. W. I.

CONTENTS

CONTENTS

CHAPTER VII

Stock and commercial sizes, lumber, mill work, sash, glass, sheet metal, wire, pipe, rope, drain tile, slate, metal roofing, ready roofing—Weight of roofing—Weights of materials—Space required for storage—Space required for farm implements—Ration for beef feeders—Table for the selection of native woods—Strength of timbers, tables—Concrete, table of proportions for different uses—Silos and silage, table—Sunshine table—Dairy score card—Kitchen score card—Estimating, cubic estimates, other approximate methods, detailed estimates, taking off quantities, units of measurement, present prices—Heating, lighting, ventilation and sewage disposal—Blue printing—Problems.

CHAPTER VIII

Books on allied subjects—Government bulletins—State Experiment Station and Agricultural College bulletins—Trade publications.

AGRICULTURAL DRAWING

CHAPTER I

INTRODUCTORY

There are two general methods of describing things, one by using words, spoken or written, the other by drawing pictures. The first is ordinary language, the second method is often called the universal graphical language.

picture from his own imagination; and the stronger his visualizing power the fuller and more interesting picture does he have. But probably no two persons reading the same story ever see exactly the same picture. In fact some very able people are almost lacking

Fig. 1.—A pen-and-ink perspective drawing.

Some writers are so skillful in their use of words, that when describing some scene or event their writing is called a word-picture, and the interested reader feels in imagination that he can see it all vividly before him. But he has simply taken the author's suggestions and has filled in the details of the

in the power of mental imagery, and in reading a story do not construct any imaginative picture at all.

It would evidently be almost impossible to describe the appearance and construction of a proposed new machine or building so that it could be built, by using words alone.

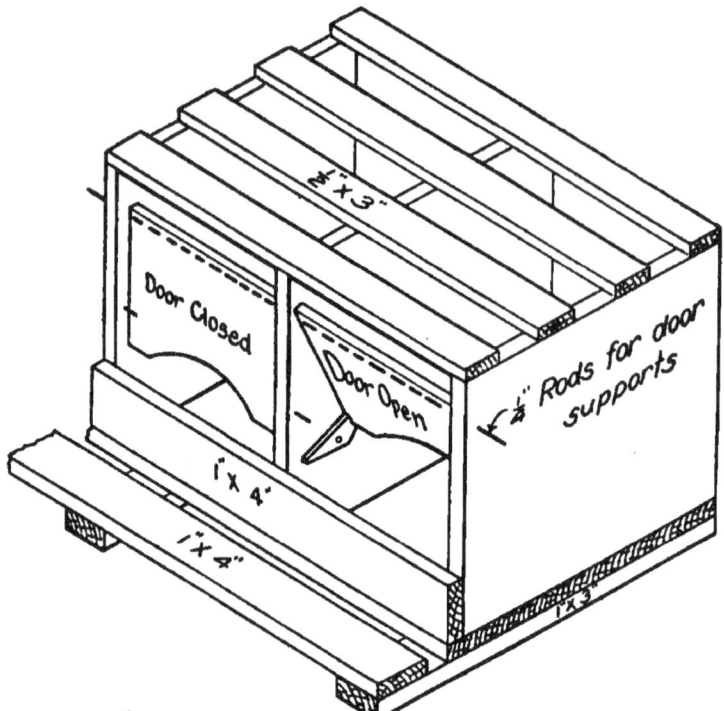

FIG. 2.—An isometric drawing.

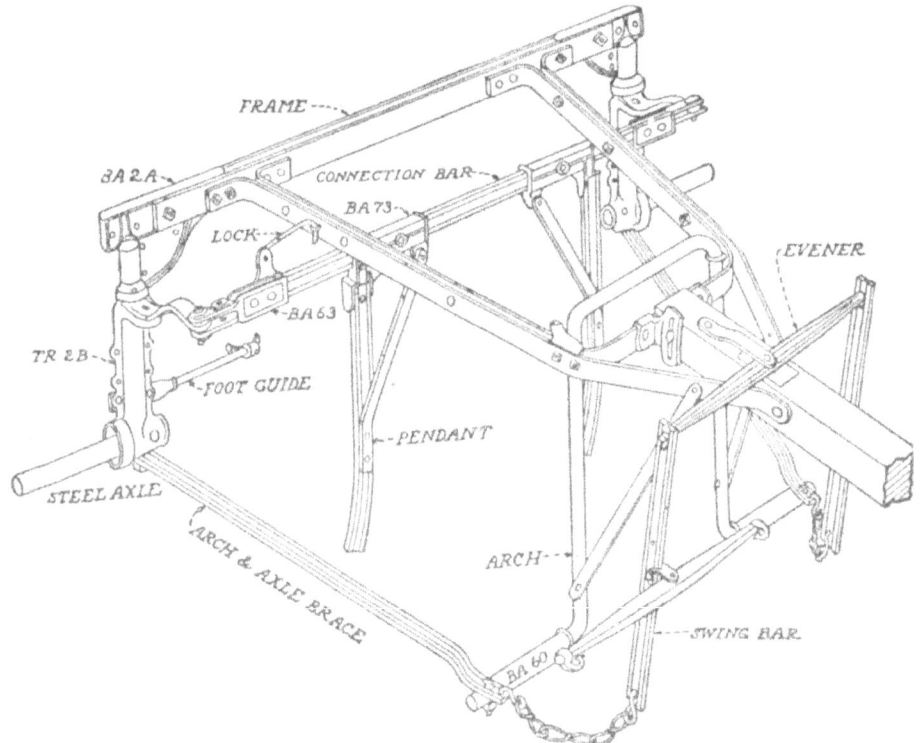

FIG. 3.—An isometric illustration.

Thus in technical description, where nothing can be left to the imagination, the second method, drawing, becomes by far the more important. The shape of even the simplest object can be explained much more accurately and quickly by a drawing than by verbal description.

But this descriptive drawing may not be simply like an artist's picture, because the artist's method of drawing is again only suggestive, and while it shows what the object looks like, it leaves much to be supplied by the observer's imagination. Technical drawing must describe accurately every

able to think in space, and an increasing power to represent and explain what he has in his mind.

DIFFERENT KINDS OF DRAWING

There are several different kinds of drawing used in technical work. They may be divided broadly into *pictorial drawings* and *working drawings*.

Pictorial Drawings.

Drawing an object as it actually appears to the eye is called *perspective drawing*. This is used by architects in showing the

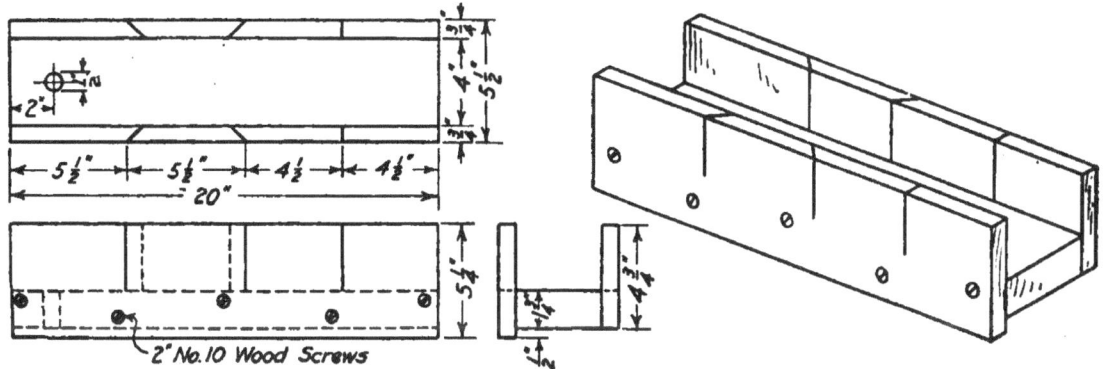

Fig. 4.—A working drawing and a picture.

detail of the structure. (The ordinary contractor has no imagination that will supply something not shown on the drawings, without being paid an "extra" for it!)

Thus this method of describing objects by lines becomes a real language, to be studied in the same way as any other language. Its thorough mastery is necessary for the professional engineer and architect, but everyone who has anything to do with building or machinery should know the elementary principles of the subject, that is, should be able to read and write in the language.

It is not a complicated nor mysterious subject, but is very simple in its principles, and the ability to learn it does not depend upon any natural talent. As one studies it he feels a growing consciousness of being

appearance of a proposed building or group of buildings. Artists and illustrators draw in perspective directly from the object or landscape before them. The architect in drawing a building not yet erected, assumes the observer to be standing at a certain point and works out the perspective from the plans by somewhat complicated methods, finishing it (or *rendering* as he calls it) in water color, pen and ink or pencil. This drawing shows the building just as a photograph or sketch taken from the same point would do; but as it cannot be measured, the drawing is of no value to build from. Perspective drawing is, moreover, too involved and difficult to be of general use. Fig. 1 illustrates the appearance of a perspective drawing rendered in pen and ink.

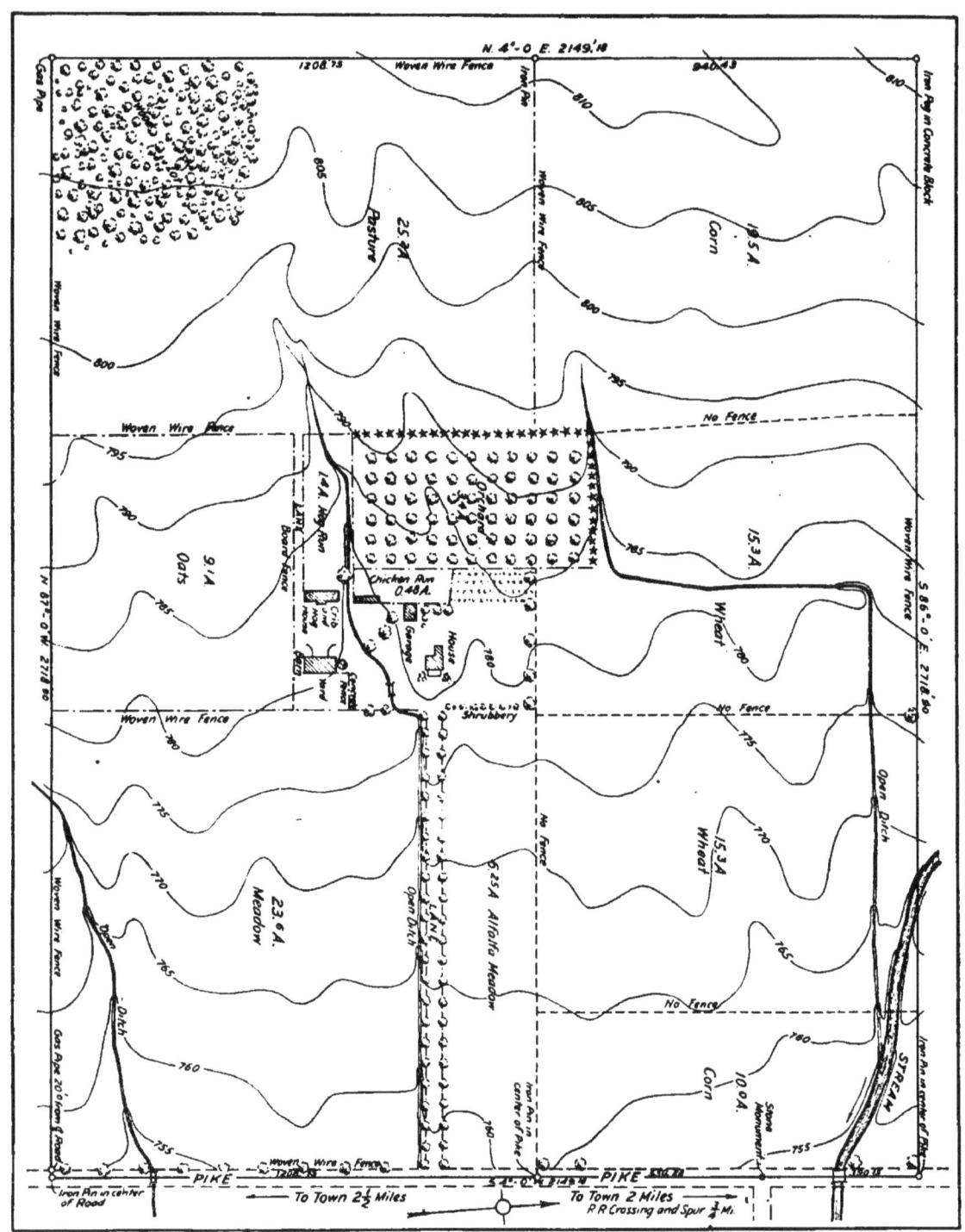

Fig. 5.—A topographical drawing.

A simpler way of making pictorial draw-ings suitable for some purposes is by iso-metric or oblique drawing. These systems are very useful in showing details of con-struction, and are also used very commonly for illustrations in bulletins and books. They are not difficult to make, and the method is fully explained in Chapter VI. Fig. 2 is a typical isometric illustration from a Government bulletin, and Fig. 3 a more complicated example as used in an imple-ment catalogue, and which shows the con-struction more clearly than any other kind of drawing would do.

Working Drawings.

A working drawing is a drawing that gives all the information for the complete con-struction of the object represented.—Thus the essential difference between a pictorial drawing and a working drawing is that the pictorial drawing shows the object as it appears, while the working drawing must show it as it actually is, giving the exact shape and dimensions of every part. This is done by making different "views" of the object, in a system known technically as "orthographic projection," the principles of which are explained in the next chapter. This is the basis of all industrial drawing, mechanical and architectural, and may be called the *grammar* of this graphical language.

Architectural drawings are working draw-ings, as they are used to build from. The architect makes perspective drawings, pre-viously referred to, principally to show his clients, who are unable to read the working drawings, what the building is going to look like. Fig. 4 illustrates a pictorial drawing and a working drawing of a simple object.

Topographic Drawing.

Still another kind of drawing, with the rudiments of which we should be familiar, is topographic drawing. This includes the drawing of maps and plats, showing the method of representing land and water features, and is the kind of drawing used in connection with surveying. Its particular value and interest to the farmer is explained in Chapter V.

Fig. 5 is a topographic drawing of a farm and farmstead, and shows not only the location of the various features, as buildings, fields, ditches, etc., but also the contour, or as it is sometimes called, the lay of the land.

CHAPTER II

THEORY AND TECHNIQUE

Of the several kinds of drawing just referred to, each has a particular use. For constructive drawings, that is for drawings of things which are to be made, the system known as orthographic projection is used almost exclusively, as being the method best adapted for showing an object exactly as it is to be. As it does not show the object as it will appear to the eye, one must be trained in reading it, and in exercising the visualizing power of the imagination to see the object from its projections.

It is our purpose in this chapter to study the principles of this system and the technique of its execution.

Constructive drawing is sometimes done freehand, when it is called "technical sketching," but for designing structures and making accurate working drawings it is necessary to use instruments and work to dimensions. Preliminary studies and schemes for new structures or machines are usually sketched first freehand, and often the final working drawing of a simple object wanted is made without instruments. Our first work in studying the principles will be done as freehand sketching, after which the technique of instrumental or mechanical drawing will be taken up.

INSTRUMENTS

For instrumental drawing an outfit of drawing materials must be at hand. Professional draftsmen use expensive, high-grade instruments. Those who will do only occasional drawing can get along with a comparatively inexpensive outfit.

The following is a list of instruments and materials needed:

1. A set of drawing instruments, containing a 6 inch compass with pen, pencil and lengthening bar, a pair of dividers, a ruling pen and a bow pen. A bow pencil and bow spacer are desirable but not necessary additions.
2. A drawing board of soft pine, cleated to prevent warping (or a drawing table with pine top).
3. A T-square. These come in various lengths and grades. For the work outlined in this book a 30″ blade is long enough.
4. A 45° triangle and a 30°–60° triangle. Transparent amber is the best.
5. An *architect's* scale, either one triangular shape or two flat ones. (An *engineer's* scale of decimal parts is needed in map work, and a protractor should also be at hand.)
6. Thumb tacks.
7. Drawing pencils.
8. Waterproof drawing ink.
9. Eraser.
10. Sandpaper pad for sharpening pencils.
11. Drawing paper. Most working drawings are made on buff or cream-colored "detail paper" which is sold by the yard, in rolls, and are afterwards traced on tracing cloth or tracing paper.

ORTHOGRAPHIC PROJECTION

Orthographic projection is the theoretical name given to a method of drawing two or more views of an object in order to show the exact shapes of its parts and their relation to each other. It means practically that we draw one view of the object as it would appear if we looked directly down on the top of it, another view looking straight at the front, and if necessary, a third looking directly at the end or side of the object. Thus if we were asked to make the projec-

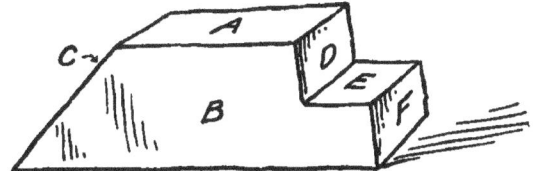

FIG. 6.—Pictorial view of block.

tions of the block shown in pictorial view in Fig. 6, we would have, by looking straight down at it from the top, a *top view* as shown in Fig. 7(a), on which the faces A, C and E would show. Looking at it directly from the front would give the *front view* (b), on which the face B only would be visible, while on the *side view* (c) the faces D and F only would show. Evidently the width of

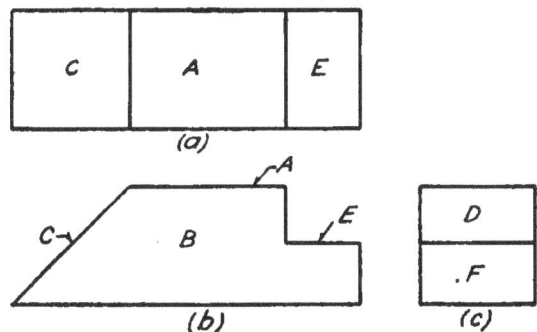

FIG. 7.—Three views of block.

the side view is the same as the width of the top view. Similarly the lengths of the front and top views are equal, and the heights of the front and side views are equal.

These three views describe completely the form of the object. In combination **the top view is always placed directly above the front view, and the side view directly across from the front view.**

Fig. 8 is the pictorial view of a bracket, and Fig. 9 the three views of the same bracket. Study and compare these views.

Explaining, perhaps more accurately and carefully, the theory of orthographic pro-

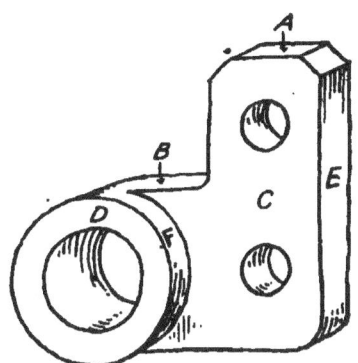

FIG. 8.—Pictorial view of bracket.

jection is that the object (Fig. 10) is conceived to be surrounded by transparent planes perpendicular to each other (as if it were inside of a box with glass sides). If lines perpendicular to these planes be extended or "projected" to them from every

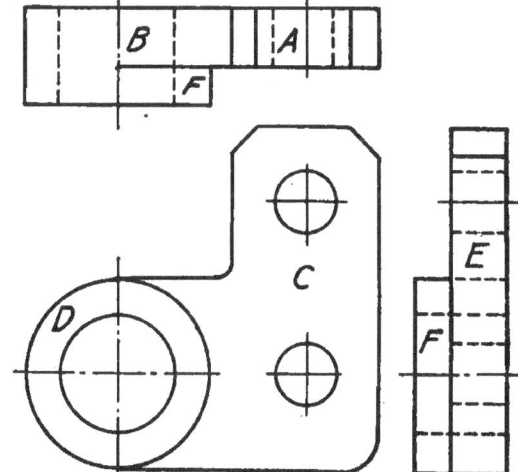

FIG. 9.—Three views of bracket.

point of the object, the resulting figures on the planes will be the "projections" of the object, Fig. 11. These planes are then imagined to be opened up into one surface, represented by the drawing paper (as if the sides of the glass box were hinged, and opened out). Thus the flat view of these opened planes would be like Fig. 12, and, leaving off the outlines of the edges, and the lines of the hinges, the three views would appear as in Fig. 13.

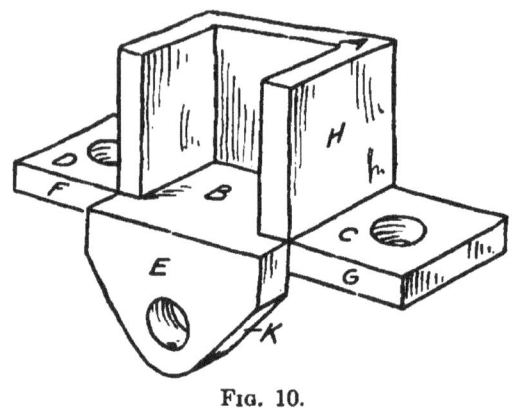

FIG. 10.

This shows the reason for the rule already given, that the top view is directly above the front view, and the side view directly across from the front view. Notice particularly that on the side view the front of the object is *facing* the front view. Sometimes two side views are necessary, and in compara-

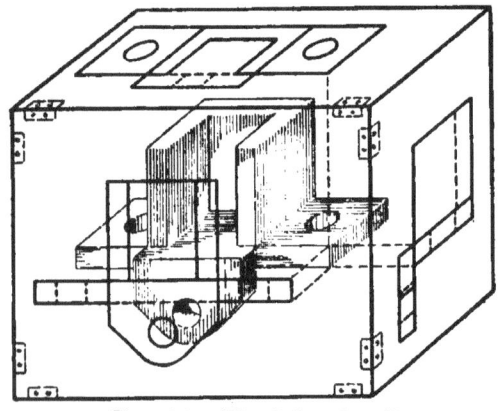

FIG. 11.—The "glass box."

tively rare cases a bottom view is desirable (as if the other end of the glass box, and the bottom of it were opened on their hinges until flush with the front).

From a study of these projections the following principles will be noted.

1. A surface parallel to a plane of projection is shown in its true size; as *B* on the front view of Fig. 7.
2. A surface perpendicular to a plane of projection is projected as a line; as

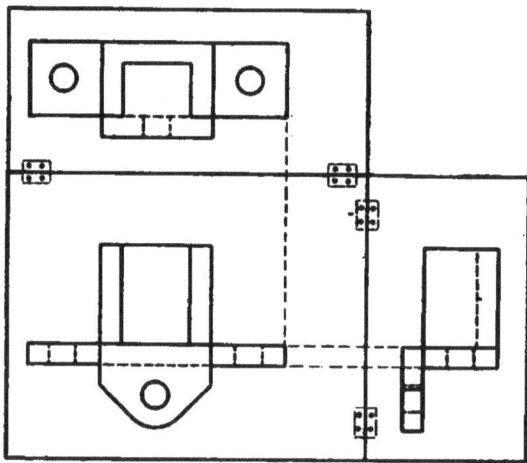

FIG. 12.—The box opened.

faces *A*, *C* and *E* on the same view (Fig. 7b).

3. A surface inclined to a plane of projection is foreshortened; as face *C* on the top view of Fig. 7, and *K* on the side view of Fig. 13.

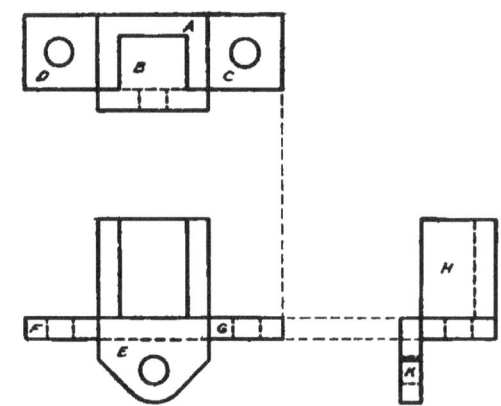

FIG. 13.—The three projections.

Similarly—

4. A line parallel to a plane of projection will show in its true length.
5. A line perpendicular to a plane of projection will be projected as a point.

6. An inclined line will have a projection shorter than its true length.

As a general rule to be followed, the view showing the characteristic contour or shape of a piece should be drawn first; thus in Figs. 6 and 8 the front view would be the

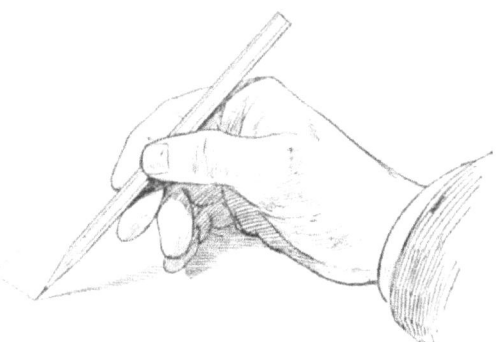

FIG. 14.—Sketching a vertical line.

first one to be made, while in Fig. 21o on page 13 the top view would be drawn first, to advantage.

In architectural drawing the top view is always called the *plan*, and the front and side views the *front elevation* and *side elevation* respectively.

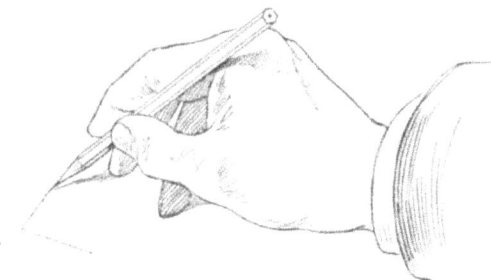

FIG. 15.—Sketching a horizontal line.

Sketching.

Training in freehand sketching is so important that there should be much practice in it. Our first work will be sketching in orthographic projection from pictorial views and models.

An H, F, or No. 3 drawing pencil sharpened to a long conical point, not too sharp, a pencil eraser, to be used sparingly, and paper, either in note book, pad, or single sheet

clipped on a board, are all the materials needed. Sometimes coordinate paper, ruled in faint lines, is used.

The pencil should be held with freedom, not close to the point, vertical lines drawn downward, Fig. 14, and horizontal lines from left to right, Fig. 15.

In beginning a sketch, after studying the object until the views are clear, mentally, always start with center lines or base lines, and remember that the view showing the contour or characteristic shape should

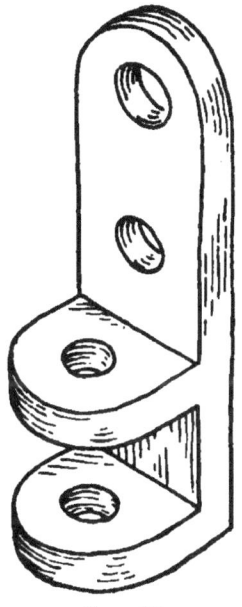

FIG. 16.

be drawn first. This is generally the view showing circles if there are any. Get the main dimensions and proportions first, blocking in the necessary number of views so that they will fit the sheet.

In such a figure as Fig. 16 the sketch would be started by drawing a vertical center line, on which would be spaced the principal points of the front and top views. The side view would then be blocked in, and the sketch at this stage would be something like Fig. 17. The holes would then be added and the outlines brightened, so that the finished sketch would look like Fig. 18.

Figs. 19, 20 and 21 are collections of pictorial views of various familiar objects, which are to be sketched in orthographic projection, making the views necessary to describe the object fully and clearly. This is primarily for practice in projection but is also a test of the student's judgment in observing and recording proportions.

exercises given in this chapter careful attention should be paid to the explanations and hints on the methods of handling the different instruments. It is very easy to get into bad habits in using the instruments unless good form be observed at the start. These habits once formed are very difficult to overcome.

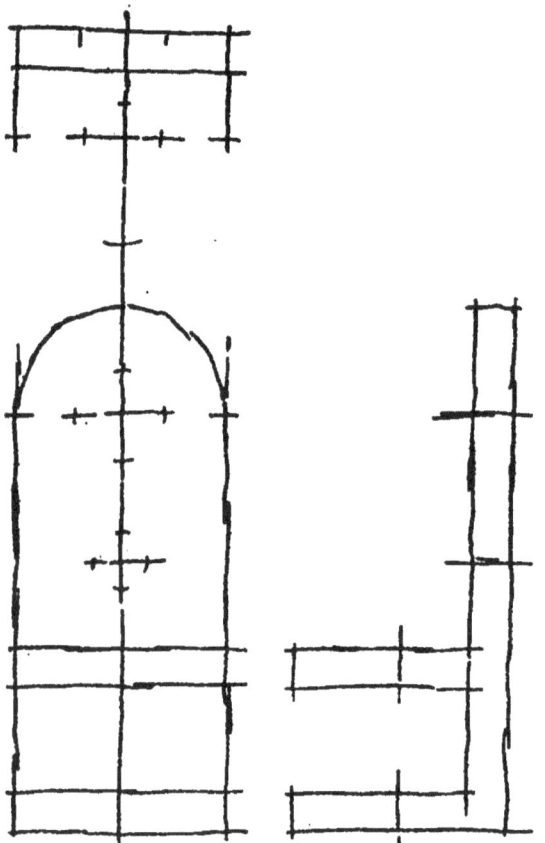

Fig. 17.—First stage of sketch.

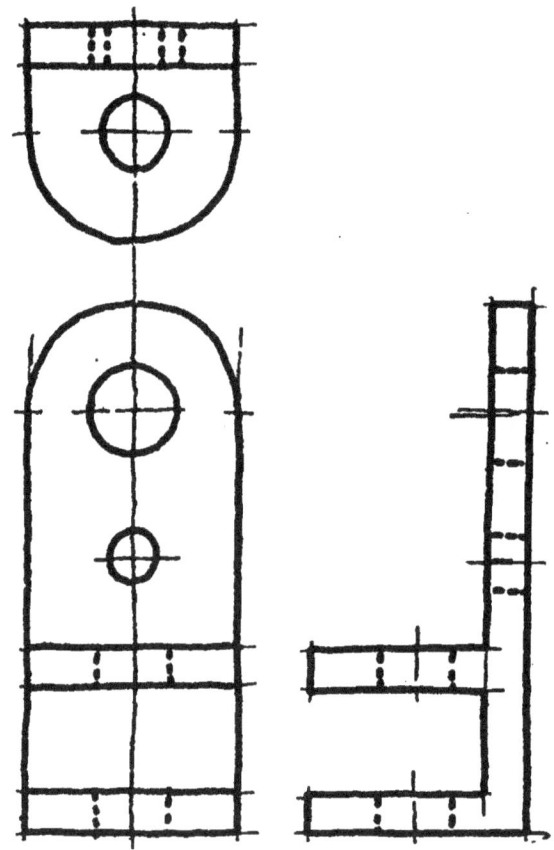

Fig. 18.—Finished sketch.

Drawing with Instruments.

As has been said, instrumental drawing is necessary in all accurate work in designing and drafting, and the first requirement is the ability to use the drawing instruments correctly. With continued practice will come a facility in their use which will free the mind from any thought of the means of expression.

Our further work in the subject will be done chiefly with instruments, and in the

Alphabet of Lines.

As the basis of the language of drawing is the line, a set of different kinds of lines needed may properly be called an alphabet of lines. The ones in general use are given in Fig. 22.

The weight of line for the visible outline will vary with the kind of drawing. Architectural and structural drawings are made with a line not heavier than shown at (1),

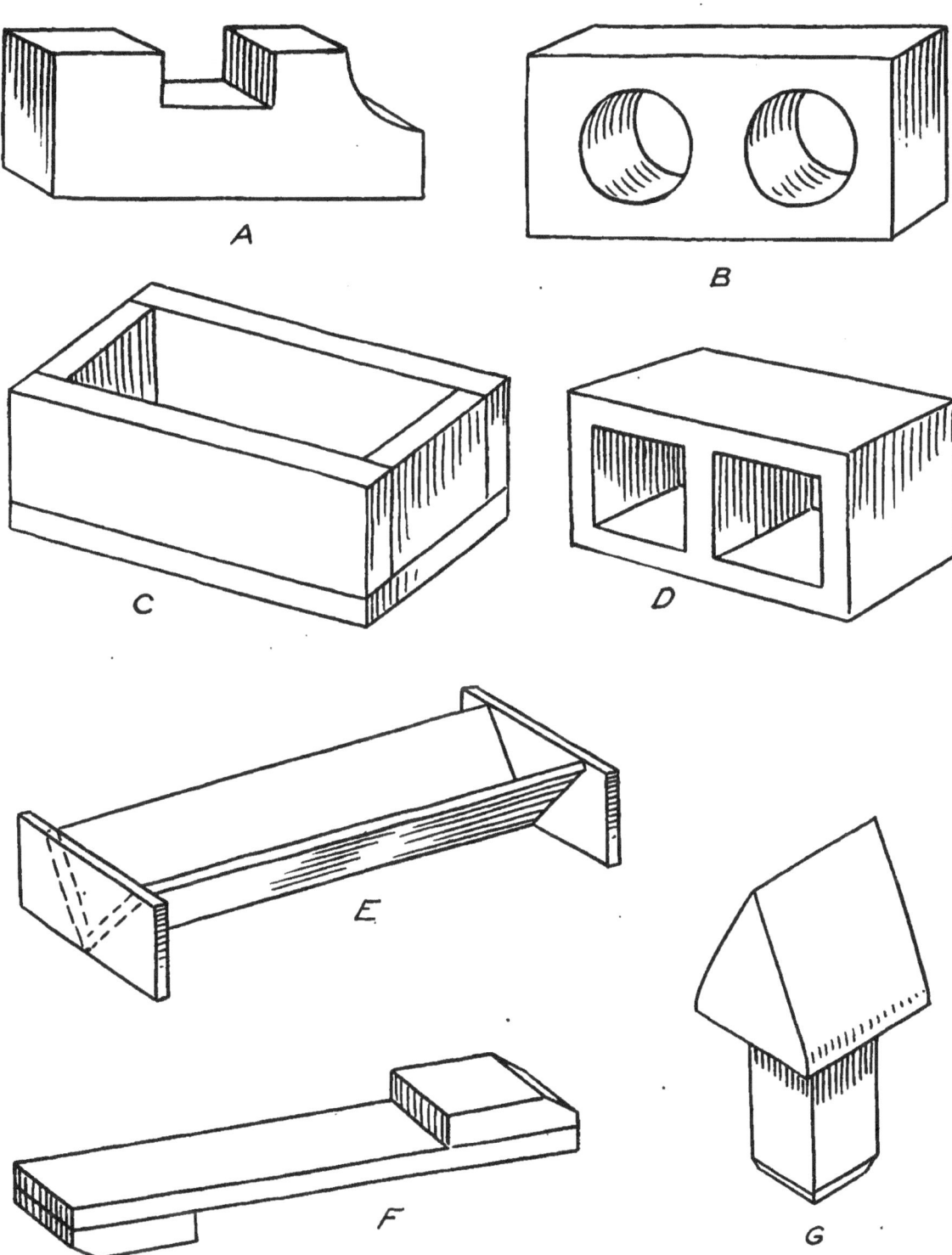

FIG. 19.—Problems for working sketches.

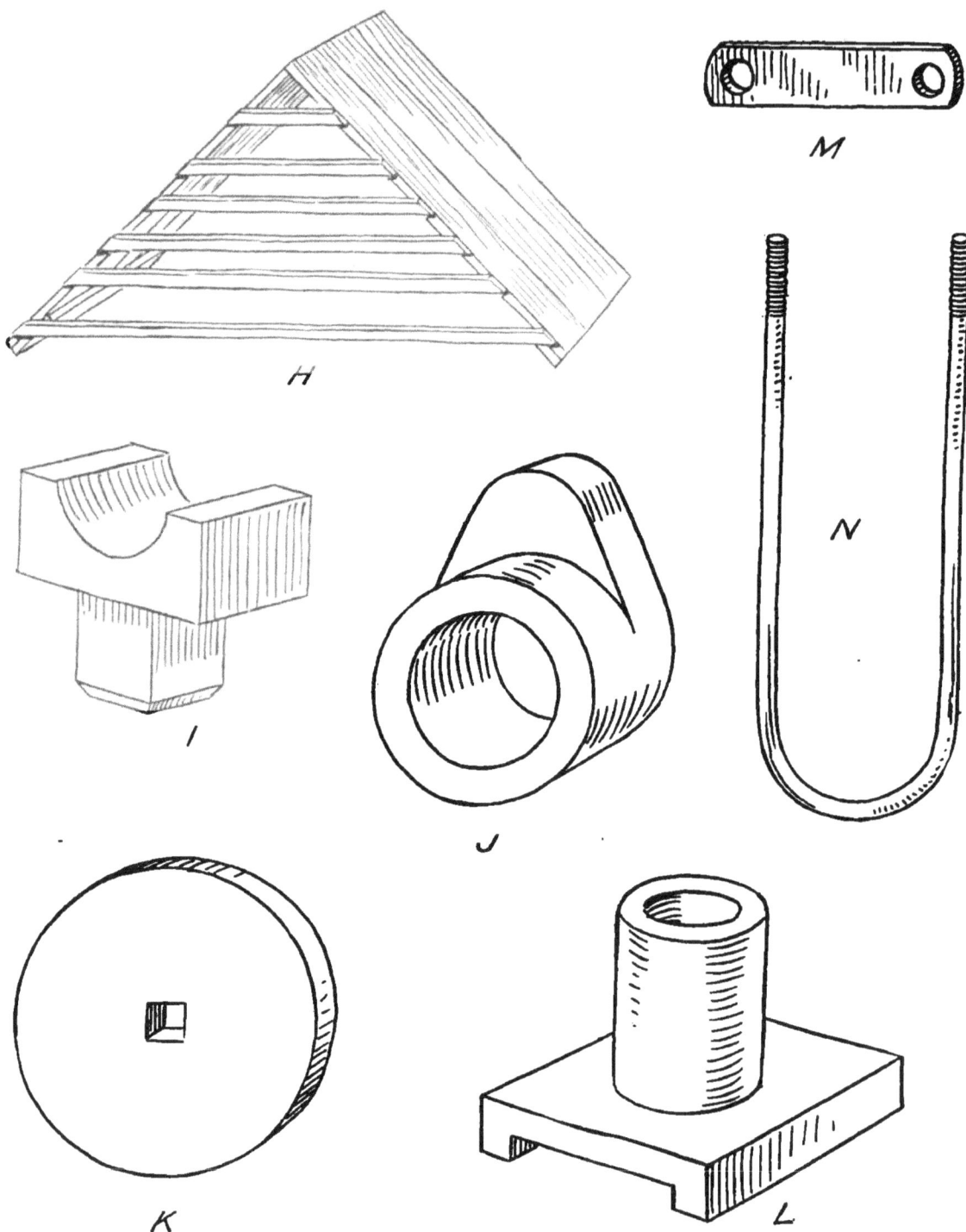

Fig. 20.—Problems for working sketches.

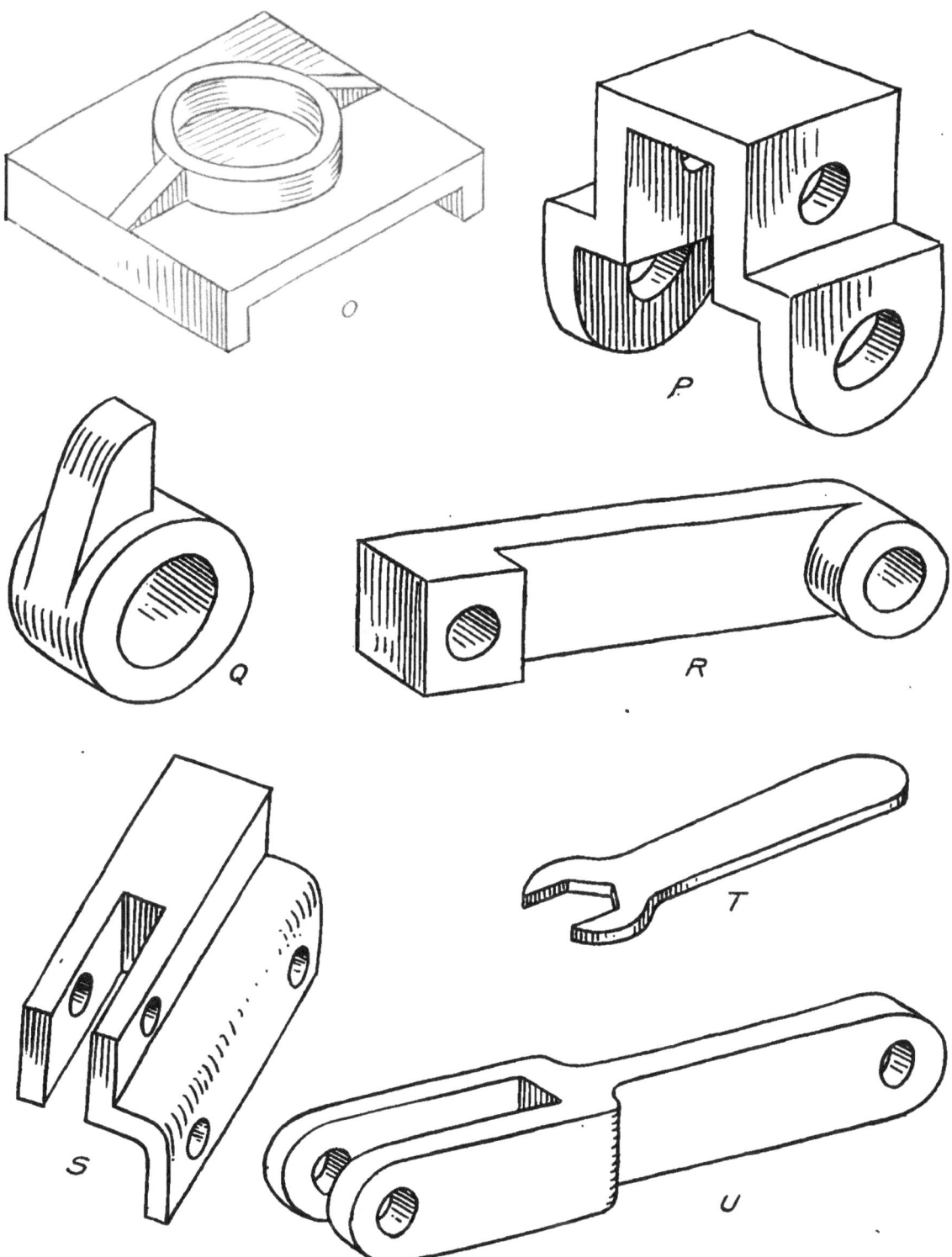

Fig. 21.—Problems for working sketches.

while drawings of machine parts are usually outlined with a much wider line.

Center lines, dimension lines and cross

Use of T Square and Triangles.

The T square is for drawing parallel horizontal lines, and is always used with its head

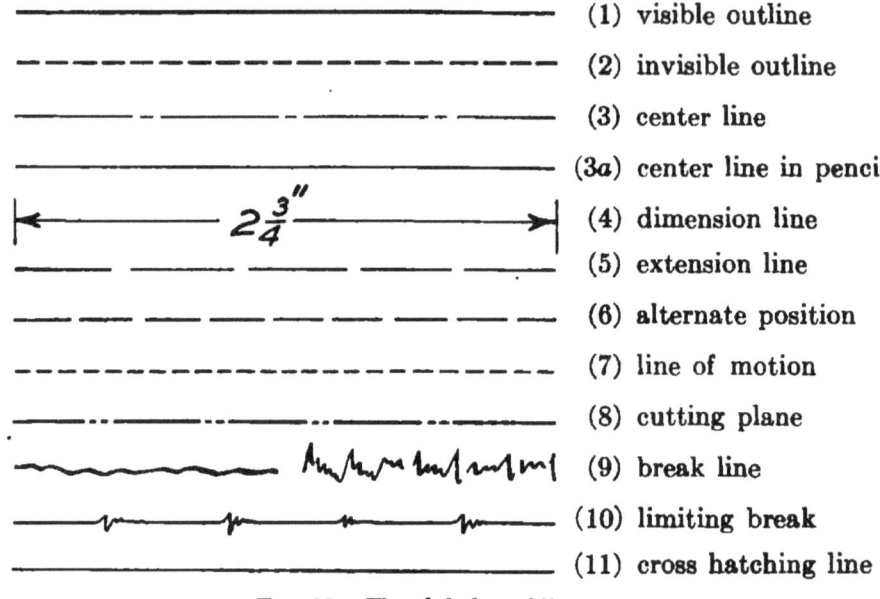

(1) visible outline
(2) invisible outline
(3) center line
(3a) center line in pencil
(4) dimension line
(5) extension line
(6) alternate position
(7) line of motion
(8) cutting plane
(9) break line
(10) limiting break
(11) cross hatching line

Fig. 22.—The alphabet of lines.

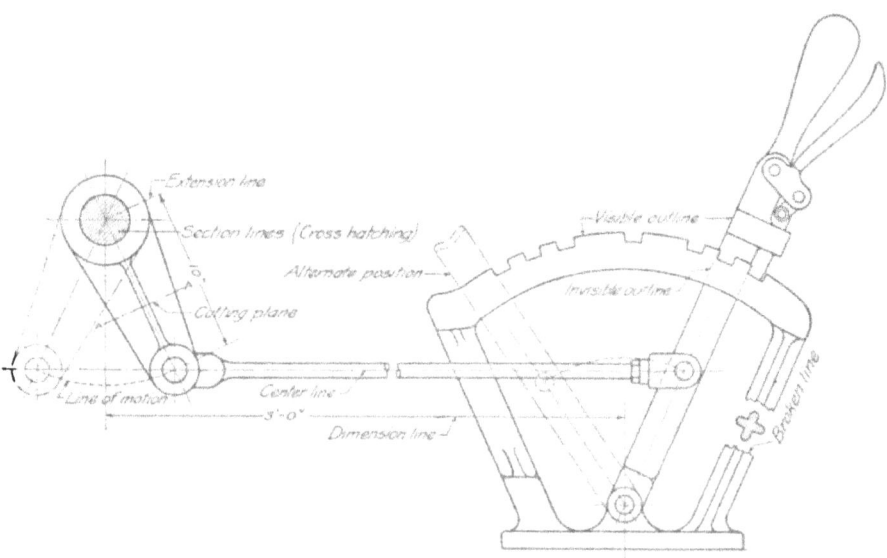

Fig. 23.—The alphabet illustrated.

hatching lines should be clean fine uniform lines. Fig. 23 shows these and the various other lines as used on a drawing.

against the *left edge* of the drawing board. The triangles are used against the T square for drawing vertical lines and lines at 30, 45 and 60 degrees.

With the two triangles together, lines at 15 degrees and 75 degrees may be drawn. Fig. 24 illustrates these combinations.

edge (the paper should preferably be a little larger than the finished drawing is to be). Lay the scale down on the paper close to the

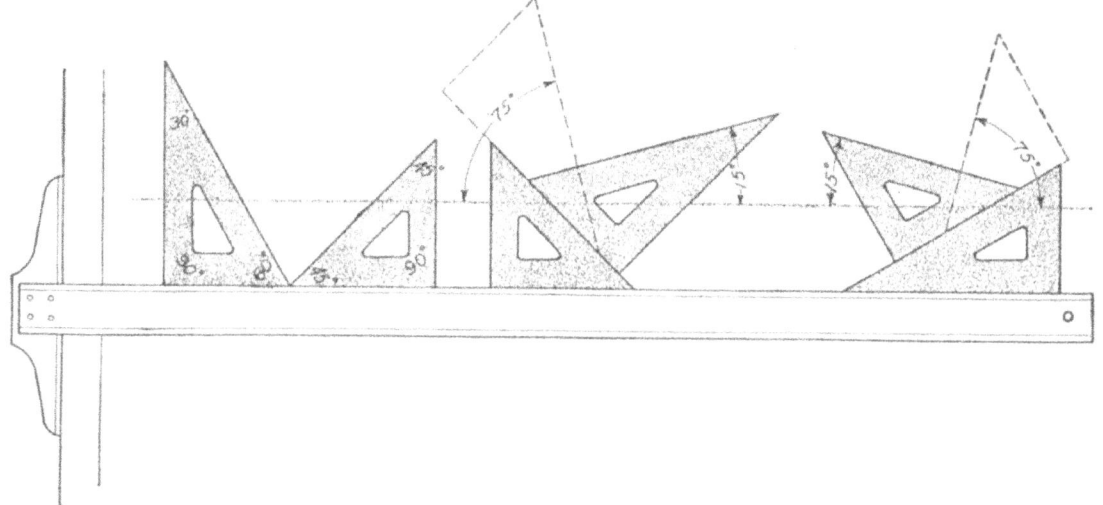

FIG. 24.—Use of triangles.

The detailed construction of the two following sheets is given to illustrate the method of procedure in making a drawing, and the use of some of the instruments. If these drawings are made following the instructions carefully, making them over if necessary until a satisfactory result is obtained, the use of the instruments and the method of laying out a drawing should be sufficiently familiar to the beginner, so that succeeding work can be done without hesitation or fault in execution.

Sharpen a hard pencil (4H or 6H) to a long sharp point, cutting away the wood and pointing the lead by rubbing it on the sand-paper pad. (For straight line work some prefer a flat wedge-shaped point, as it stays sharp longer.) Have the sand-paper pad always at hand and **keep the pencil point sharp.**

Fasten a piece of paper to the board, squaring the top edge with the T square, and putting a thumb tack in each corner, pushing the tacks down to their heads.

Suppose the size of the drawing is to be 12″ × 18″ with a border line ½″ from the

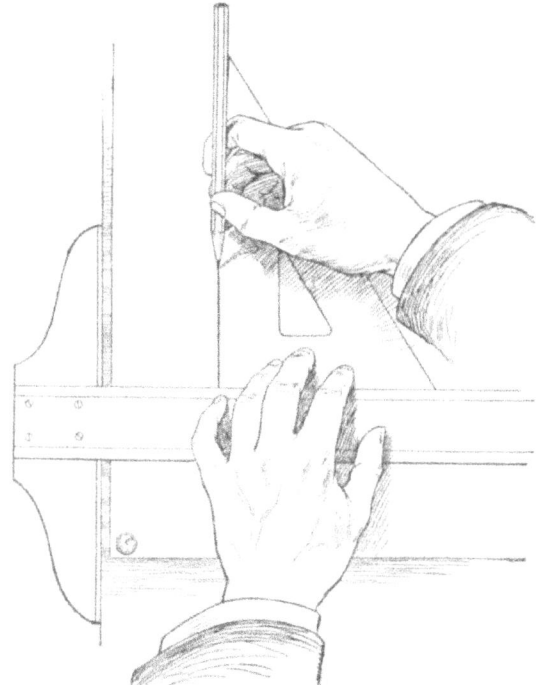

FIG. 25.—Drawing a vertical line.

lower edge and measure 18″, marking the distances with the pencil, at the same time marking ½″ for the border line. Always use a short dash, not a dot, in laying off a dimension.   At the left edge mark 12″ and ½″ border line points.   Through these four points on the left edge, draw horizontal lines with the T square, and through the points

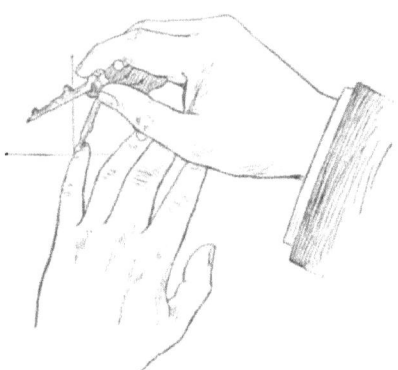

Fig. 26.—Setting the compass.

on the lower edge draw vertical lines with the triangle against the T square in the position illustrated in Fig. 25.   Horizontal lines should always be drawn from left to right, and vertical lines upward.

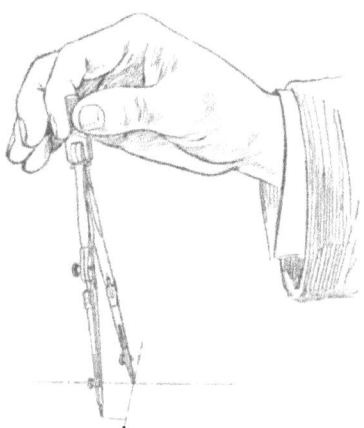

Fig. 27.—Starting the circle.

Use of the Compasses.

In drawing a circle the radius should be measured and marked on the center line and the compass adjusted to it by first pinching the instrument open with the thumb and second finger, then setting the needle point (shoulder point) in position at the center and

adjusting the pencil point to the mark, using one hand only in opening and closing the compasses.   The needle point may be guided to the center with the little finger of the left hand, Fig. 26.   When the lead is

Fig. 28.—Completing the circle.

adjusted to pass exactly through the mark, the hand should be raised to the handle and the circle drawn (clockwise) in one sweep by twirling the handle with the thumb and forefinger, keeping the compasses inclined

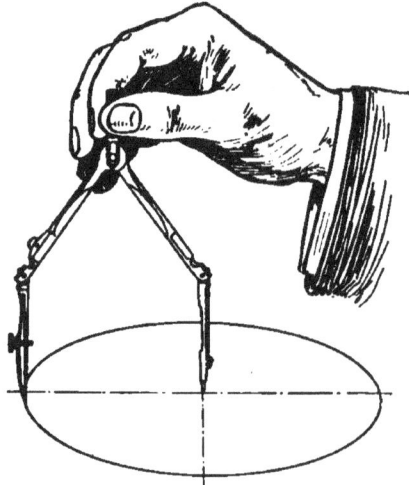

Fig. 29.—Position for large circle.

slightly, Fig. 27.   The position of the fingers after completing the circle is shown in Fig. 28.   Circles up to perhaps 3 inches in diameter may be drawn with legs straight but for larger sizes the legs are bent at the

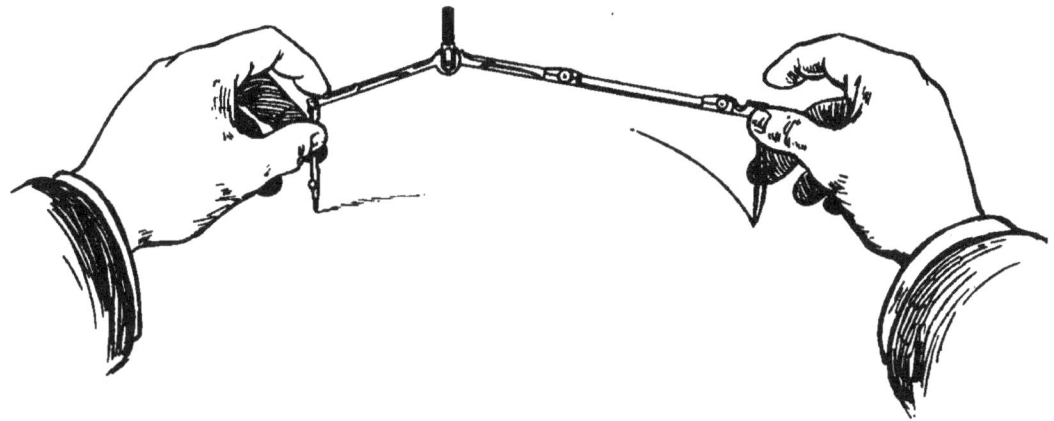

Fig. 30.—Use of lengthening bar.

joints so as to be perpendicular to the paper, Fig. 29. Circles too large for the reach of the compass are drawn with

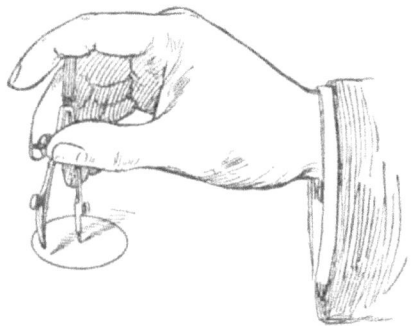

Fig. 31.—Setting the bow pen.

the lengthening bar as shown in Fig. 30. Small circles, particularly when there are a number of the same diameter, are made

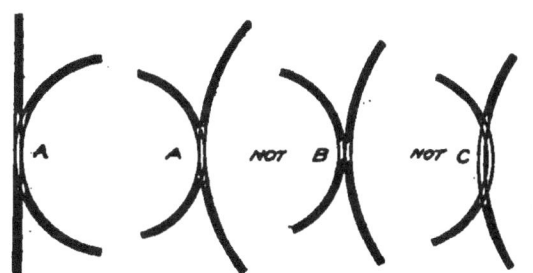

Fig. 32.—Correct and incorrect tangents.

with the bow-pen. In changing the setting, to avoid wear and final stripping of the thread, the pressure of the spring against the nut should be relieved by holding the points in the left hand and spinning the nut in or out with the finger. Small adjustments are made with one hand, with the needle point in position on the paper, Fig. 31. Notice particularly, in drawing tangent circles, that two lines are tangent to each other when their centers are tangent, and not when the lines simply touch each other; as illustrated in enlarged form in Fig. 32.

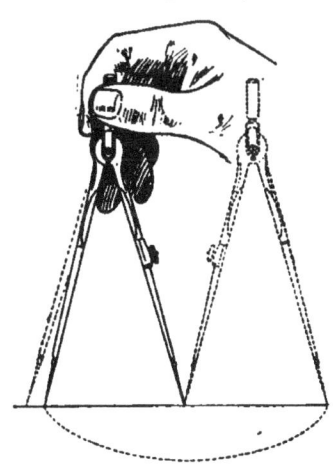

Fig. 33.—Bisecting a line with dividers.

Use of Dividers.

The dividers are used for transferring measurements from one part of the drawing to another, for stepping off distances, and for dividing lines by trial. They are handled in the same way as the compasses. Fig. 33

illustrates the method of bisecting a line by trial, first opening the dividers at a guess to one-half the length of the line and stepping the distance off. If the division be short, the leg should be thrown out to one-half the remainder, estimated by the eye, without removing the other leg from its position on the paper, and the line spaced again with the new setting.

Avoid pricking unsightly holes in the paper. The position of a small prick point may be preserved if necessary by drawing a little ring around it with the pencil.

between front and side views, and subtracting these 15″ from the length of the sheet, 17″, we find that the left edge of the front view should be started an inch from the left border.

Now with the scale measure along the base line the horizontal dimensions for the front view, marking points for the thickness of the end boards. At the same time measure on the base line the width of the side view, leaving an inch between views. Through the first point on the front view draw a long perpendicular, measure on this the height of the front view and the thickness of the bottom board. On the same line mark the width of the top view, leaving about 1¼″ between views. Through these points draw horizontal and vertical lines, thus blocking out the three views. Next draw on the side

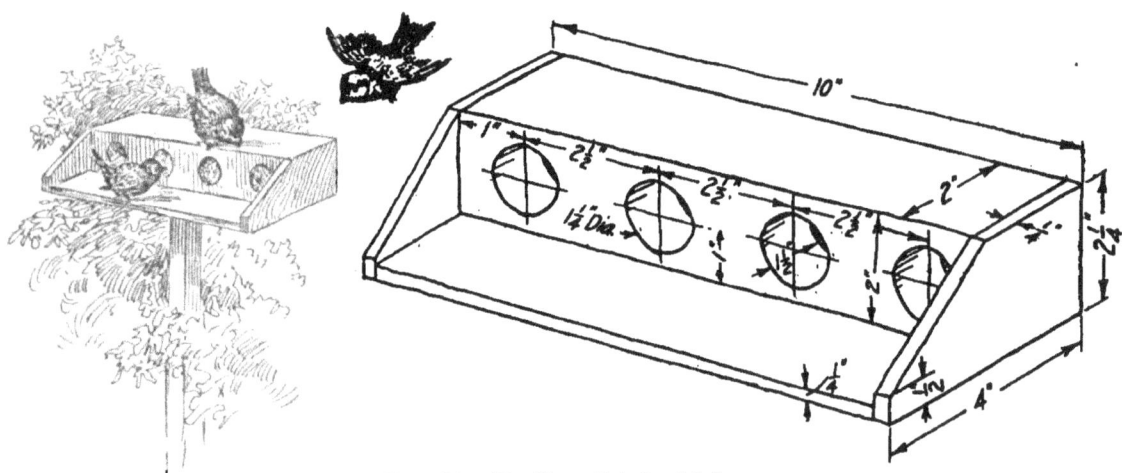

Fig. 34.—Feeding stick for birds.

PROBLEM 1.

The first sheet is to be three views of the feeding stick[1] for birds shown in Fig. 34.

Lay off a 12″ × 18″ sheet with ½″ border as described on page 15.

The first requirement of a good drawing, after deciding on the requisite views, is to have the views well spaced on the sheet. A quick preliminary freehand sketch will aid in this study. In adding the width of the top view, 4″, and the height of the front view, 2¼″, and leaving say 1¼″ between views, we would have 7½″. Subtracting this from the width of the sheet inside the border, 11″, and dividing this space between top and bottom, we find that the base line should be drawn about 1¾″ from the bottom border line. Adding the length, 10″, and the width, 4″, and allowing an inch

view, the slant line of the end boards and a dotted line showing the block behind. This block will show as a full line on the top view. On the front view draw the center line A–B for the holes, and on this measure the distances for the centers, and at one of the centers mark the radius (¾″) of the circle. At this stage the drawing will appear something like Fig. 35.

When the circles are drawn and the short line formed by the chamfer of the end boards is projected across from the side view, the front view is completed.

The top and side views of the holes, showing their depth, are hidden lines, shown as indicated in the alphabet of lines, and are projected from the front view.

When the drawing is finished in pencil it is to be inked. It is shown in finished stage in Fig. 36.

Over-running pencil lines should not be erased until after the drawing is inked.

[1] Such a feeding stick, with the holes filled with suet is greatly enjoyed by birds in winter.

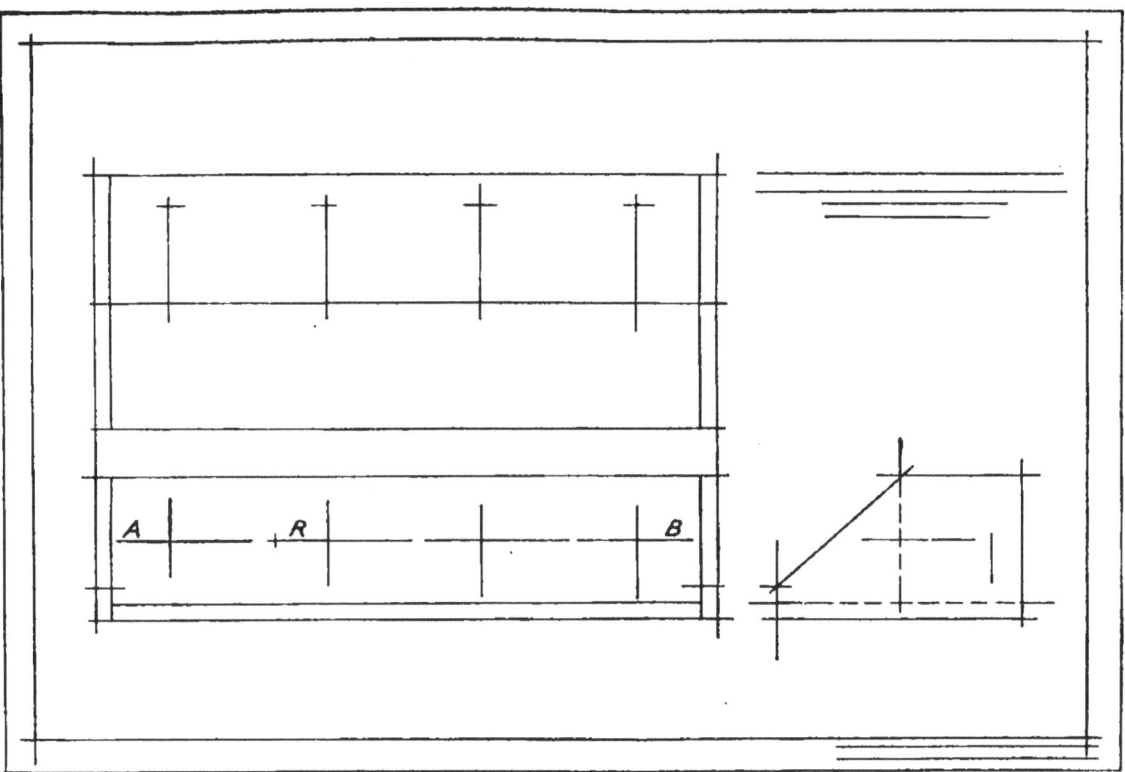

FIG. 35.—First stage of penciling.

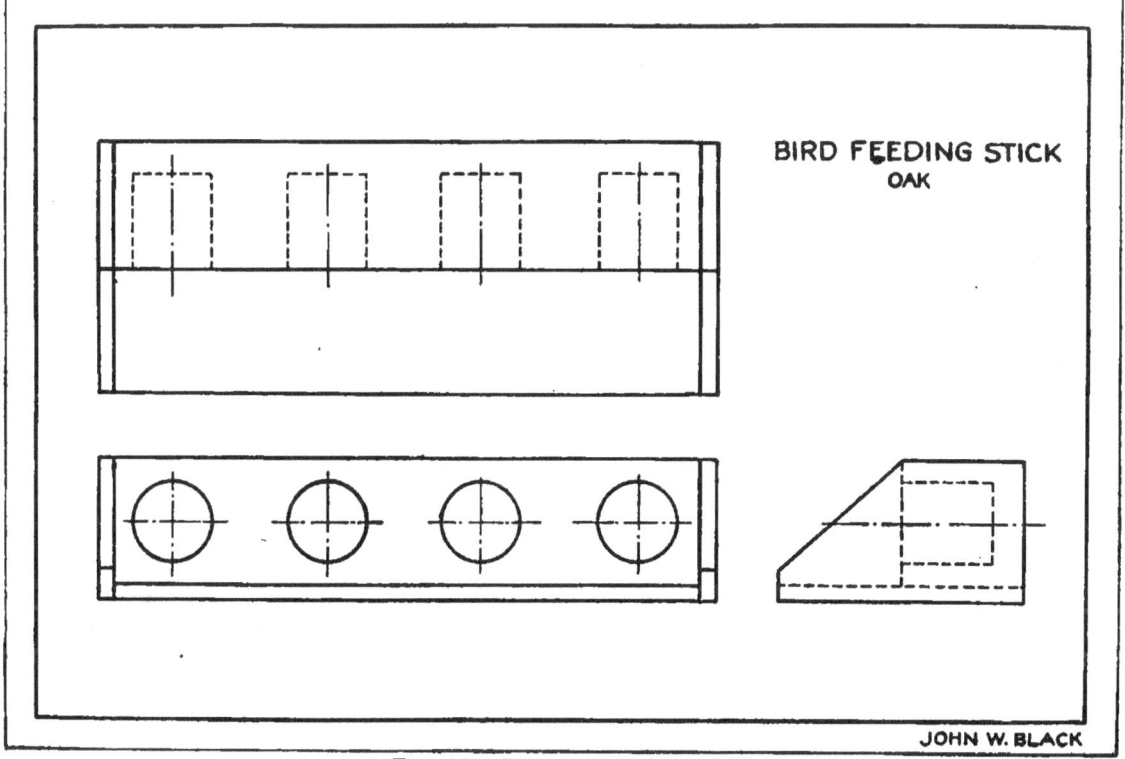

FIG. 36.—Finished ink drawing.

Inking.

Finished drawings are either inked on the paper or traced on tracing cloth, or sometimes on tracing paper. Straight lines are inked with the ruling pen, which is filled by

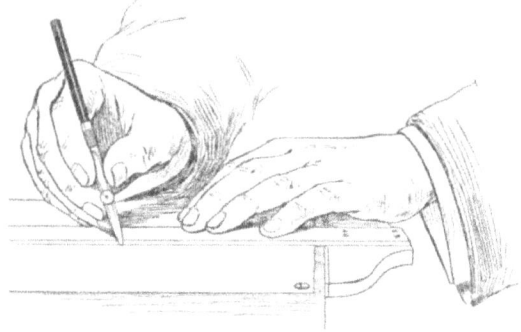

FIG. 37.—Correct pisition of ruling pen.

touching the quill filler attached to the cork of the ink bottle, between the nibs of the pen, being careful not to get any ink on the outside of the blades. Not more than three-sixteenths of an inch should be put in or it

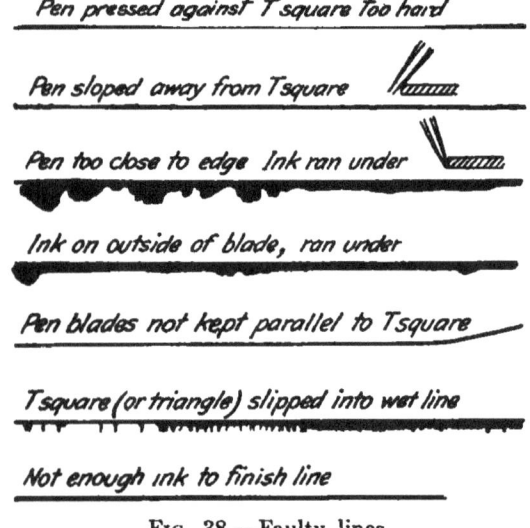

Pen pressed against T square too hard

Pen sloped away from Tsquare

Pen too close to edge Ink ran under

Ink on outside of blade, ran under

Pen blades not kept parallel to Tsquare

Tsquare (or triangle) slipped into wet line

Not enough ink to finish line

FIG. 38.—Faulty lines.

will drop out in a blot. After adjusting the nibs with the screw to give the correct thickness of line, the pen is held as illustrated in Fig. 37. Keep it in a plane perpendicular to the paper and draw the lines with T square and triangle.

If the ink refuses to flow it is because it is dried and clogged in the extreme point of the pen. This clot or obstruction may be removed by touching the pen on the finger or by pinching the blades slightly. If it still refuses to start it should be wiped out and fresh ink added. The pen should always be wiped clean after using.

Faulty Lines.

If inked lines appear imperfect in any way, the reason should be ascertained immediately. Fig. 38 illustrates the characteristic appearance of several kinds of faulty lines. The correction in each case will suggest itself.

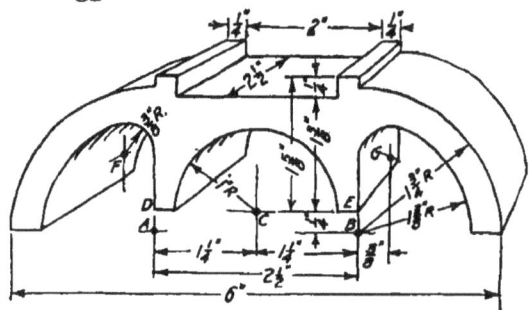

FIG. 39.—Windmill brake shoe.

PROBLEM 2.

The next sheet, three views of a windmill brake shoe, Fig. 39, is an exercise in straight and curved lines.

Draw the border lines, figure the spacing as was done on Sheet 1 and lay out base line, and a vertical center line for the front view. Evidently the front view is the one to make first. On the base line of the front view set off the centers A and B $1\frac{1}{4}''$ from the vertical center line. Measure up $\frac{1}{4}''$ on the vertical center line and draw the horizontal line DE. On this mark the $1''$ radius for the middle circle, whose center is at C. Measure up from the center C $1\frac{3}{8}''$ and draw the upper horizontal line. Next draw vertical lines through A and B. With the radii indicated draw the circle arcs with A, B and C as centers. The construction for the arcs from the centers F and G is suggested on the figure. Complete the measurements for the front view, then draw top and side views, measuring the widths and projecting the other dimensions from the front view. Fig. 40 illustrates a partially completed stage of the drawing, and Fig. 41 the finished drawing. Remember as a fundamental rule that

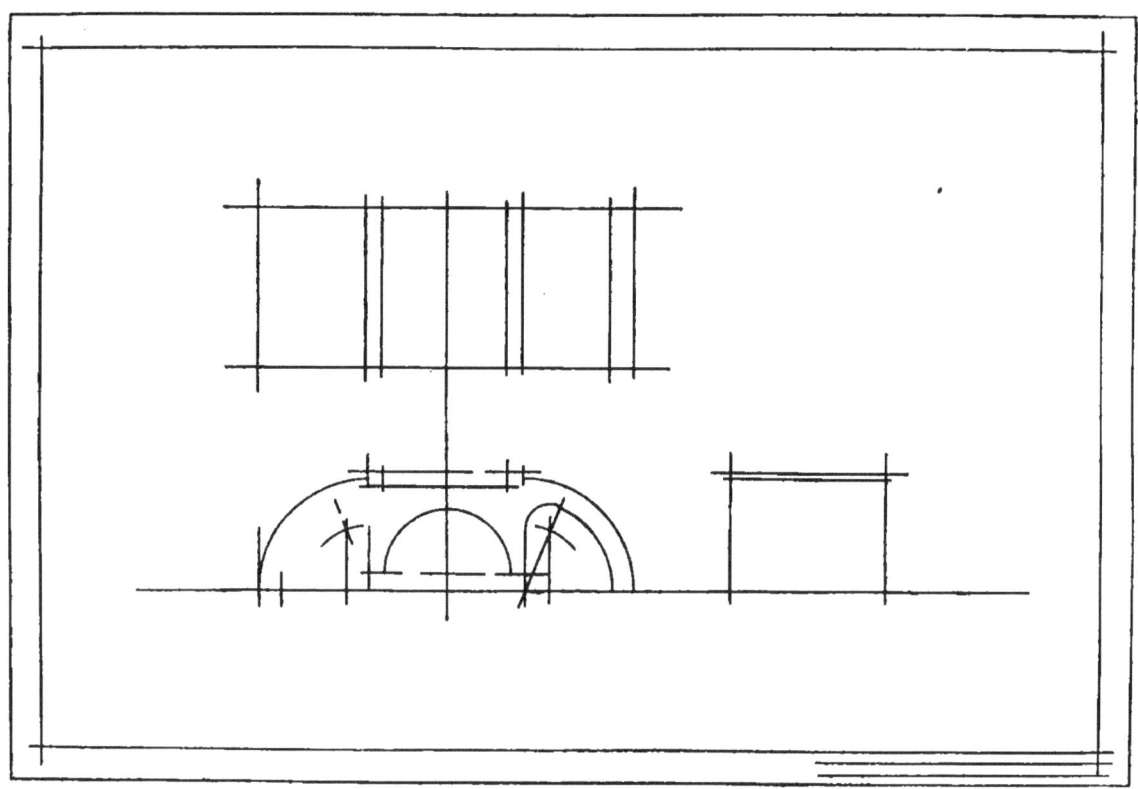

FIG. 40.—First stages in penciling.

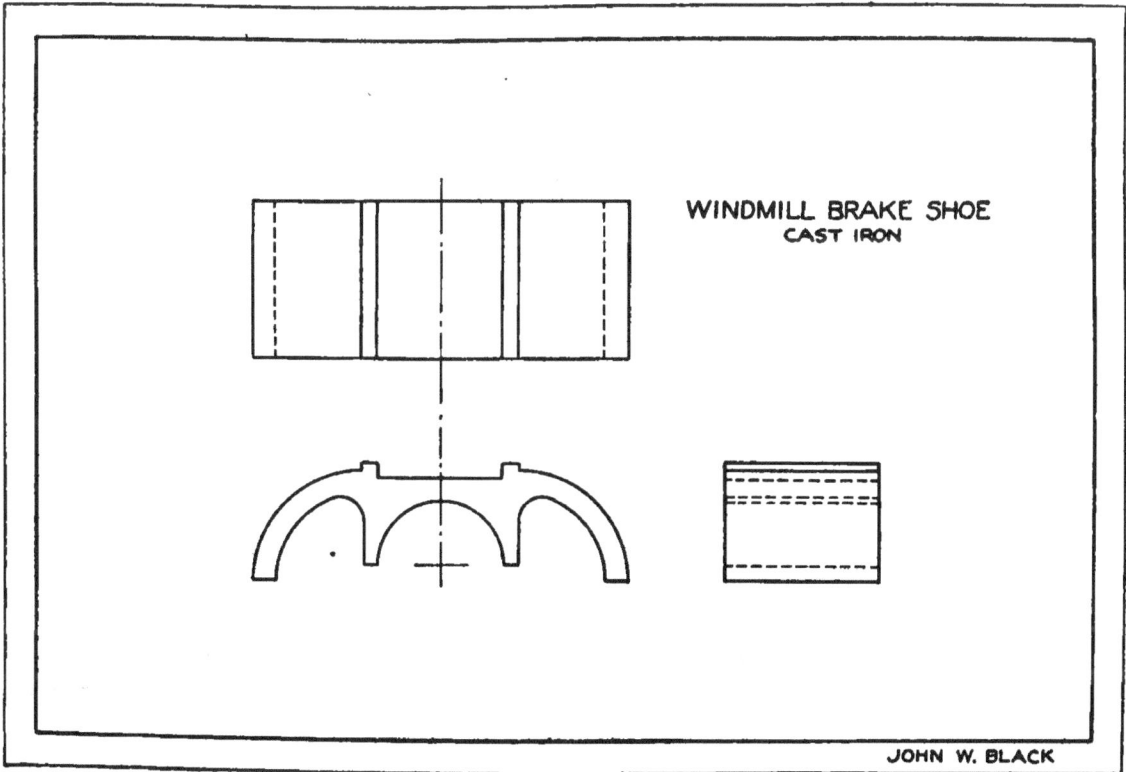

FIG. 41.—Inked drawing.

circles and circle arcs are always inked before the straight lines are inked.

LETTERING

A working drawing requires the addition of dimensions, notes on material and finish, only by continued and careful practice. Working drawings are lettered in a rapid single-stroke style, either vertical or inclined, and usually all capitals. The term "single-stroke" or "one stroke" does not mean that the entire letter is made without lifting the

ABCDEFGHIJKLMN
OPQRSTUVWXYZ&
1234567890½

FIG. 42.—Upright single-stroke capitals.

and a title, all of which must be lettered free-hand in a style that is perfectly legible, uniform and capable of rapid execution. So far as its appearance is concerned, there is no part of a drawing so important as the letter-ing. A good drawing may be ruined not only in appearance, but in usefulness by lettering done ignorantly or carelessly, as illegible figures are very apt to cause mistakes in the work.

The ability to letter well can be acquired pen, but that the width of the stroke of the pen is the width of the stem of the letter. For large sizes in this style, a comparatively coarse pen such as Hunt's No. 512 or Leonard's ball point No. 516 F is used.

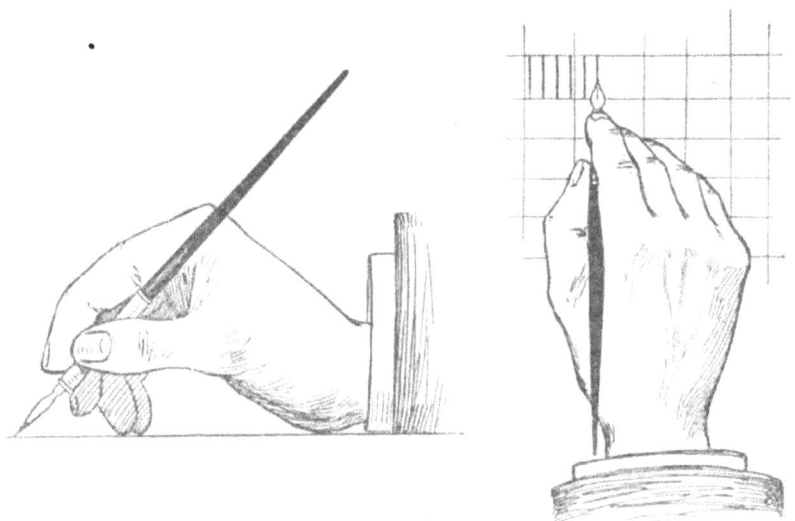

FIG. 43.—Position for lettering.

Single-stroke Vertical Caps.

The upright single stroke "commercial gothic" letter shown in Fig. 42 is one of the standard forms. To practice this letter draw a page of guide lines $\frac{3}{16}$" apart, hold the pen in position shown in Fig. 43 and practice each letter a number of times, following the order and direction of strokes given in Fig. 44, and watching the copy

Single-stroke Inclined Caps.

Many draftsmen prefer inclined letters to vertical letters. They should be practiced

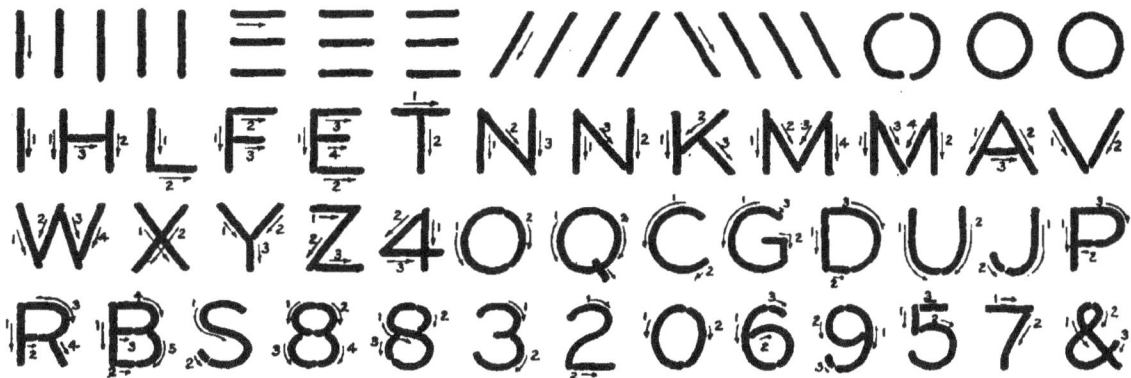

FIG. 44.—Order and direction of strokes.

carefully. Letters should then be combined into words, remembering three general rules:

1. Keep the letters close together.

in the same way as explained for vertical letters, but with the addition of slant direction lines drawn lightly in pencil with a

ABCDEFGHIJKLMN
OPQRSTUVWXYZ&
1234567890½

FIG. 45.—Inclined single-stroke capitals.

2. Have the areas of the white spaces between letters approximately equal (thus two o's would be spaced much closer together than two n's).

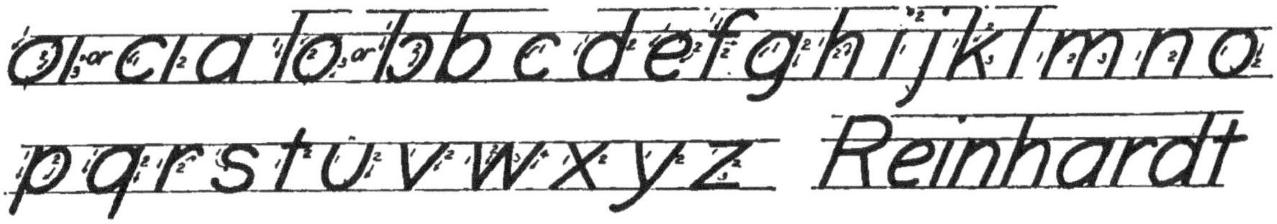

triangle either at 60 degrees or at a slant of 2 to 5, in order to keep the slant uniform. Fig. 45 shows the alphabet with the order and direction of strokes indicated.

a or c a b or b b c d e f g h i j k l m n o
p q r s t u v w x y z   Reinhardt

FIG. 46.—Single-stroke lower case, showing strokes.

3. Keep words well separated, leaving a space between them at least equal to the height of the letters.

Single-stroke Inclined Lower Case.

In connection with the inclined capitals, a standard letter for notes known to engineers as the Reinhardt letter, is much used. This is a very simple, legible and effective style which can be made very rapidly after its swing has been mastered. The body letters are made two-thirds the height of the capitals. It is shown in analyzed form in Fig. 46.

standardized in form, and examples are shown in connection with the drawings in Chapters IV and V.

Single-stroke Roman.

Fig. 47 shows a single stroke letter based on the Roman, which is appropriate for

ABCDEFGHIJKLMN
OPQRSTUVWXYZ&
1234567890½

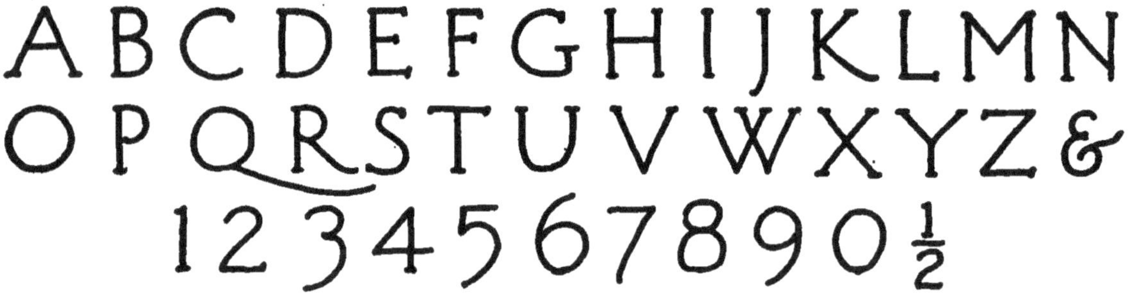

FIG. 47.—Single-stroke Old Roman.

As soon as the letter forms have been mastered, all the practice should be in composition.

Never do any lettering without having a guide line for both tops and bottoms of the letters.

Do not combine vertical letters and inclined letters on the same drawing.

Titles and bills of material are somewhat

architectural drawings. The addition of the terminal cross strokes, called serifs, requires a little more time than the straight Gothic letters, and it will be noted that the shapes vary somewhat from the Gothic. When well executed, this letter adds much to the beauty of an architectural drawing and should be used for such work in preference to the other styles given.

A PAGE OF CAUTIONS

Never use the scale as a ruler.

Never draw with the lower edge of the T-square.

Never cut paper with a knife and the edge of the T-square as a guide.

Never use the T-square as a hammer.

Never put either end of a pencil in the mouth.

Never jab the dividers into the drawing board.

Never oil the joints of compasses.

Never use the dividers as reamers or pincers or picks.

Never take dimensions by setting the dividers on the scale.

Never lay a weight on the T-square to hold it in position.

Never use a blotter on inked lines.

Never screw the nibs of the pen too tight.

Never run backward over a line either with pencil or pen.

Never leave the ink bottle uncorked.

Never hold the pen over the drawing while filling.

Never dilute ink with water. If too thick throw it away. (Ink once frozen is worthless afterward.)

Never try to use the same thumb tack holes when putting paper down the second time.

Never scrub a drawing all over with the eraser after finishing. It takes the life out of the inked lines.

Never put instruments away without cleaning. This applies with particular force to pens.

Never fold a drawing or tracing.

Never put a writing pen which has been used in ordinary writing ink, into the drawing-ink bottle.

CHAPTER III

WORKING DRAWINGS

The definition of a working drawing has already been stated as being "a drawing which gives all the information necessary for the complete construction of the object represented." It is a technical description of a structure or machine which has been designed for a certain purpose and place, and should convey all the facts regarding it so clearly and explicitly that no further instruction to the builder would be required.

The drawing will thus include, (1) the full graphic representation of the shape of every part of the object, (2) figured dimensions of all the parts, (3) explanatory notes giving specifications in regard to material, finish, etc., (4) a descriptive title, and in some cases a bill of material.

In architectural drawing the notes of explanation and information regarding details of materials and finish are often too extensive to be included on the drawings, so are written separately and are called the **specifications**. These specifications have equal importance and weight with the drawings.

The basis of practically all working drawing is orthographic projection, as explained in Chapter II. Thus to represent an object completely, at least two views would be necessary, often more. The general rule would be, make as many views as are necessary to describe the object, **and no more**.

Classes of Working Drawings.

Working drawings are divided into two general classes, *assembly drawings* and *detail drawings*. An assembly drawing or general

drawing is, as its name implies, a drawing of a structure or machine showing the relative positions of the different parts.

A detail drawing is a drawing of a separate piece or group of pieces, giving the complete description for the making of each piece. In a very simple case the assembly drawing may be made to serve also as the detail drawing, by dimensioning it fully.

Fig. 114 on page 60 illustrates an assembly drawing, and Fig. 115, page 61 a detail drawing.

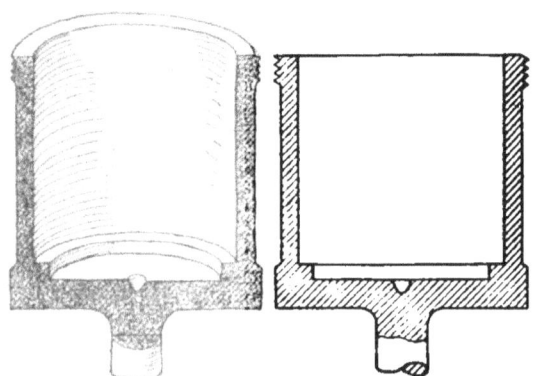

Fig. 48.—Illustration of a section.

Sectional Views.

Often it is not possible to show interior construction clearly by dotted lines on the exterior view. In such cases a view is drawn as if the object were sawed through and the front removed so as to expose the interior, as illustrated in Fig. 48. A view of this kind is known as a sectional view, or a "section," and the surfaces of the materials thus cut are indicated by "section lining" or "cross-hatching" with diagonal lines. Two adjoining surfaces are sectioned in different directions.

A line is drawn (line 8 in the alphabet of lines) on the adjacent view to show where the section is assumed to be cut, and short arrows indicate the direction in which the

Section lines are spaced by eye, and on a finished drawing are generally put in directly in ink without penciling.

"Turned sections" or revolved sections are little sectional views drawn on some part of a figure as if they had been cut and re-

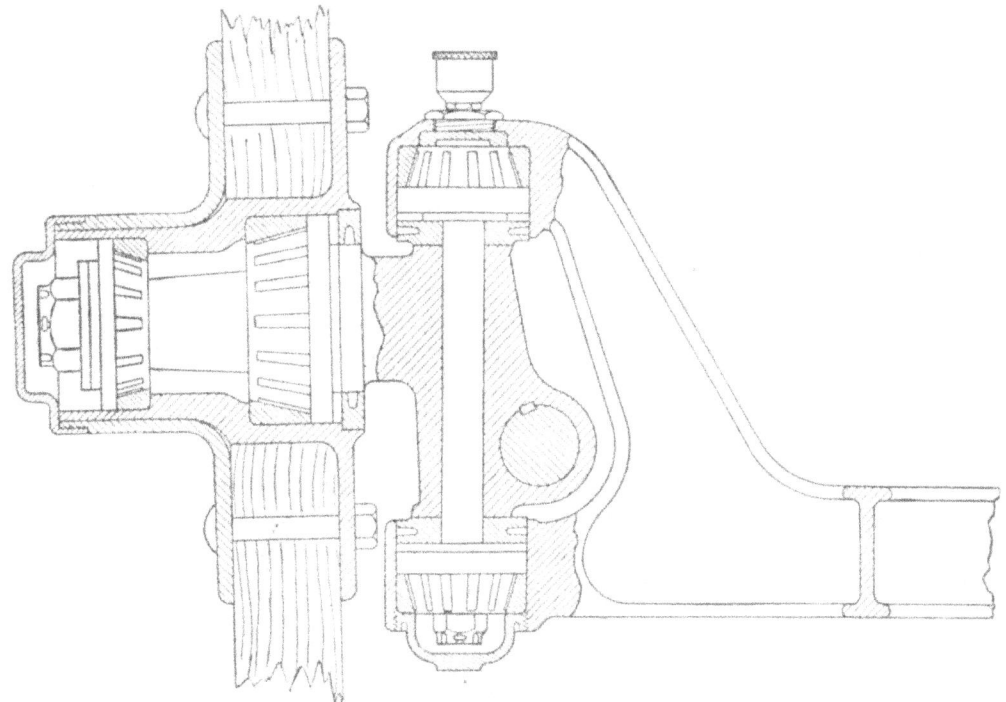

FIG. 49.—Section of roller bearing.

view is taken. This line is lettered on each end and the section is named to correspond, as "Section A——B". (See Fig. 54.) Sometimes several sections through different places are required to explain the construction fully.

A common practice, which saves space and labor, is to show one-half a view in section and the other half in full, as in Fig. 116, page 62.

In a sectional view it is not necessary to cut through everything in the plane of the section. For example, bolts, nuts, shafts, etc., show to better advantage if left in full. Fig. 49 illustrates the method of showing a section containing shafts, bearings and bolts, all of which are drawn in full.

volved in place. They show the shape of the cross-section of the object at that place, and are often used to good advantage. Examples are shown in Figs. 49 and 50.

FIG. 50.—Revolved sections.

Auxiliary Views.

Sometimes when it is necessary to show some feature on an inclined face of an object, it can be done to better advantage by making what is known as an auxiliary view, which may be thought of as simply a projection looking straight against the inclined surface, as illustrated in Fig. 50. Evidently the width of this auxiliary view would be the true width of the object, as indicated by "*W*" on the figure, and its length is of course shown in edge on the front view. Thus to draw an auxiliary view we would first draw a center line for it, parallel to the inclined

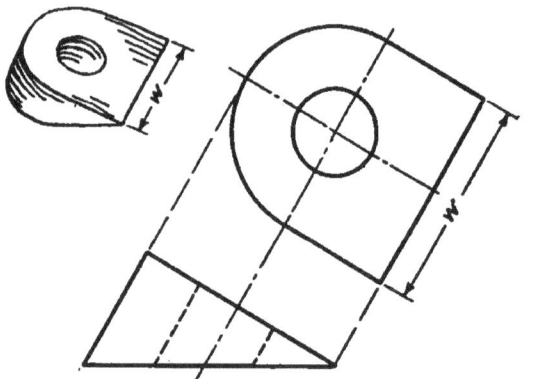

Fig. 51.—An auxiliary view.

face, then project the lengths across perpendicular to this center line from the inclined face, and would measure the width each way from the center line. Thus the width of the auxiliary view would be the same as the width of the top view of the object, although in many cases, as in the example given, a top view is not needed, and therefore not drawn.

sixteenths, which is used both for full size, and also for half size (scale of 6 inches = 1 foot), by considering the divisions on the scale to have double values.

If the object is too large to be drawn half size, the drawing is made to the scale of 3 inches to the foot, often called quarter size, that is, a length of 3 inches on the drawing is equal to one foot on the object. On this scale the distance of 3 inches is divided into twelve equal parts and each of these subdivided into eighths. This distance should be thought of not as 3 inches but as a *foot*, divided into inches and eighths of inches.

Notice that the divided foot has the zero on the inside, with the inches running from it one way, and the feet numbered the other way, so that dimensions given in feet and inches may be measured directly, as illustrated in Fig. 52. On the other end of the 3 inch scale will be found the scale of 1½ inches = 1 foot. Other scales used are 1 inch = 1 foot, ¾ inch = 1 foot, ⅜ inch = 1 foot, ¼ inch = 1 foot, 3⁄16 inch = 1', and ⅛ inch = 1 foot. The scale of ¼ inch = 1 foot is the usual one for ordinary house plans. For very large buildings the scale of ⅛ inch = 1 foot is used. Fig. 53 illustrates the scale in position for measuring the length 7 feet 5 inches.

The scale and its divisions should be

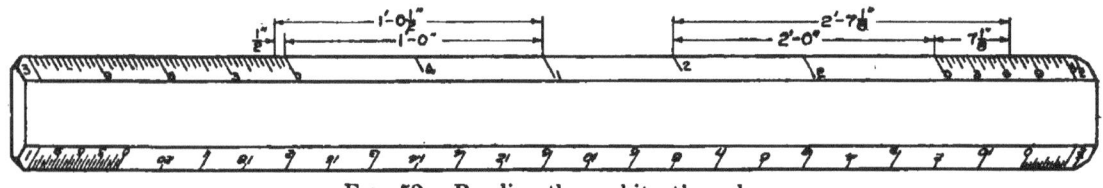

Fig. 52.—Reading the architect's scale.

Use of the Scale.

In representing objects which are larger than can be drawn full size, the dimensions are reduced proportionately by the use of the architect's scale. On the usual triangular form there are eleven different scales, including the full size scale of inches and

studied until it can be used with facility. See for example what fraction of an inch is represented by the smallest division on each scale; and notice that while the foot marks are numbered in both directions, the numbers for the smaller scale are always closer to the edge than those for the larger one.

Dimensioning.

After the correct representation of the object by its projections, the entire value of the drawing as a working drawing lies in the dimensioning. This placing of the figured dimensions must be done with careful thought of the purpose of the drawing,

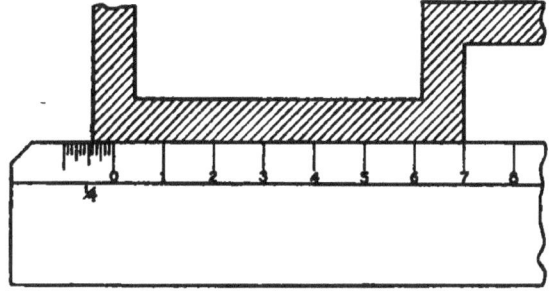

FIG. 53.—Measuring with scale (7 ft. 5 in.)

getting the figures necessary and most convenient for the workman who is to work from it. To make the best drawing the draftsman should be familiar with the various processes of construction and shop methods which enter into the building of the object represented.

General Rules for Dimensioning.

Dimension figures are placed in a space left in the dimension line, as shown in the alphabet of lines, and the exact points between which the distance is measured are indicated by arrow heads on the ends of the line, short extension lines being drawn from the object if the dimension line is on the outside.

All the extension and dimension lines should be drawn before any figures are added.

Dimensions always indicate the finished size of the piece.

Dimensions should read from the bottom and right side of the sheet, no matter what part of the sheet they are on.

Feet and inches are indicated thus 5'-6''. Where there are no inches it should be written 5'-0'', 5'-0½''.

Fractions should be made with a horizontal line as 2¼''.

The diameter of a circle should be given, not the radius.

Dimensions should generally be placed between views.

In general do not repeat dimensions on adjacent views.

Always give an over-all dimension. Never require the workman to add or subtract figures.

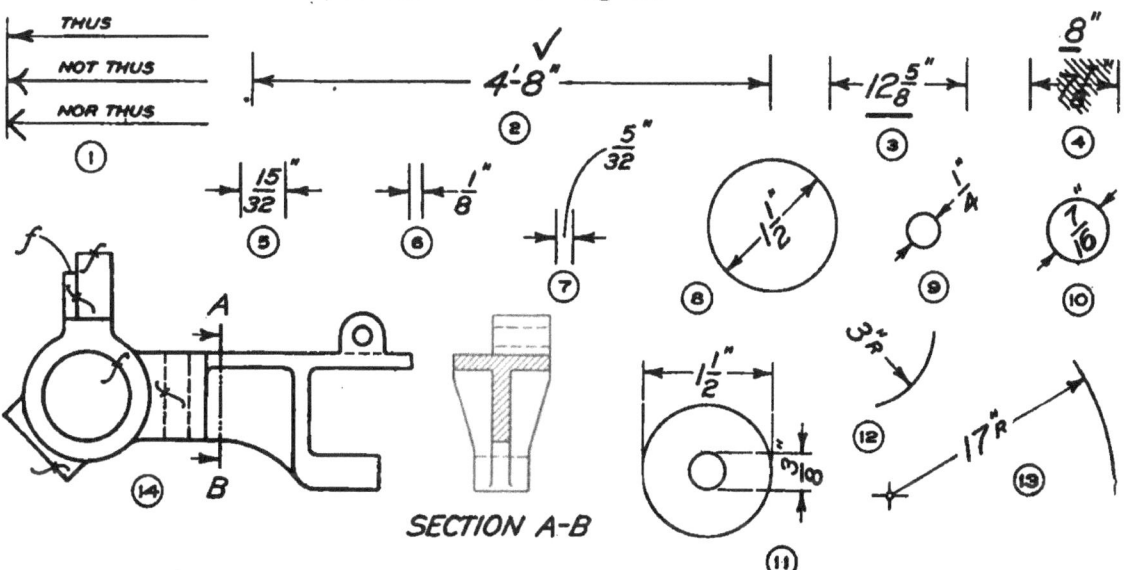

FIG. 54.—Dimensioning.

Never use any center line as a dimension line.

Never put a dimension on a line of the drawing.

A dimension not agreeing with the scaled distance, should be heavily underscored as in Fig. 54.

Fig. 54 should be observed carefully, noting the shape of the arrow heads, and the method of showing dimensions in special cases. The figure also illustrates the use of the symbol "f" to indicate that a metal

Several examples are shown in Chapter IV. For a large structure, as a barn for example, the bill of material is too extensive to be put on the drawings, so is written separately.

Checking.

Before being sent out for use, a working drawing should be checked for errors and omissions, if possible by some one other than the man who made it. In doing this the system below should be followed, taking the divisions through one at a time.

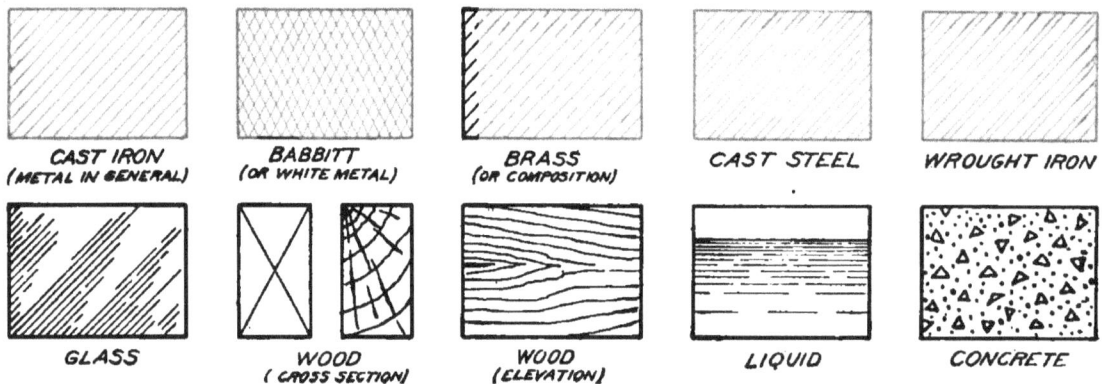

Fig. 55.—Symbols for materials in section.

surface is to be "finished" or machined, and the placing of the check mark.

Do not be afraid to put notes on drawings. Supplement the graphic language by the English language whenever added information can be conveyed, but be careful to word it so clearly that the meaning cannot possibly be misunderstood.

The **title** to a working drawing is usually boxed in the lower right hand corner, and its contents will vary according to the kind of drawing. In general it should contain the name of the structure, name of manufacturer or owner, date, scale, and drafting record including number, initials or name of draftsman, tracer, checker, etc. Various titles are shown in Chapter IV.

The **bill of material** is a tabulated form on the drawing giving the name, number wanted, size and material of each piece.

First, put yourself in the place of the one who is to work from the drawing, and see if it is easy to read.

Second, see that each piece is correctly illustrated.

Third, check all dimensions by scaling, and also by calculation where necessary. As each dimension is verified, put a check mark (√) in pencil above it.

Fourth, see that all specifications for material are correctly given.

Fifth, see that stock sizes of materials have been used as far as possible.

Sixth, add any explanatory notes that will increase the efficiency of the drawing.

CONVENTIONAL SYMBOLS

In technical drawing there are certain signs and simplified outlines called "conventions" which are adopted and recognized as standing for materials or commonly used constructions. Screw threads and gear wheels for example are not drawn in their actual outline but are shown conventionally. Other conventions are used for electrical

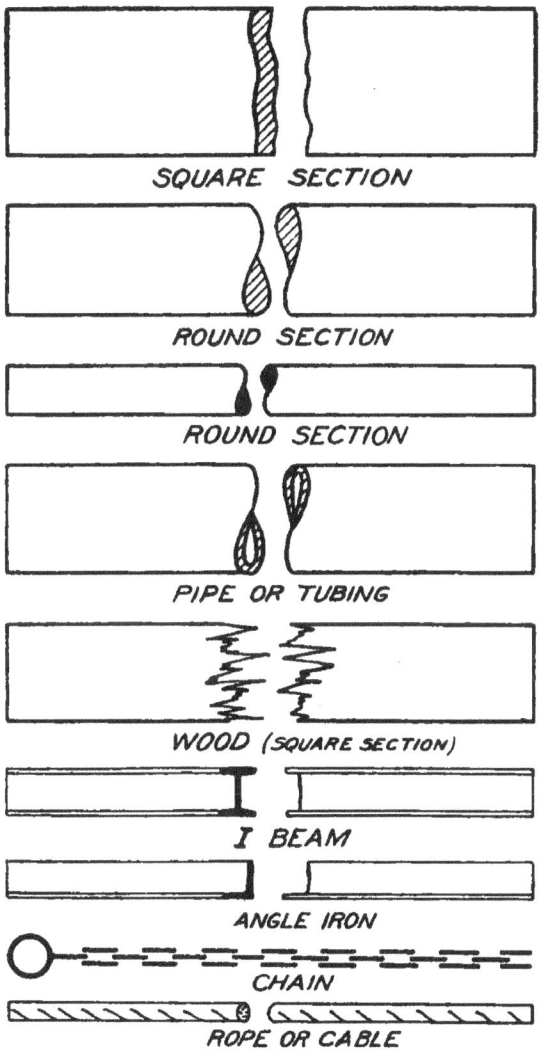

SQUARE SECTION

ROUND SECTION

ROUND SECTION

PIPE OR TUBING

WOOD (SQUARE SECTION)

I BEAM

ANGLE IRON

CHAIN

ROPE OR CABLE

FIG. 56.—Conventional breaks and symbols.

wiring and apparatus, for sections of concrete, wood, etc., and for topography. Such of these as may be useful in agricultural drawing have been given in various places in this book.

In specifying materials it is much safer to add the name of the material as a note than to depend on a symbol. The drawing is easier to read however if generally recognized sections are used for the commoner materials. Fig. 55 shows some of the conventional section lining symbols, and Fig. 56 a number of conventional breaks and other symbols. Symbols used in building construction are shown in Fig. 96, page 48.

A long object of uniform section may be shown to larger scale and thus to better advantage by drawing it as if a piece were broken out of the middle and the ends pushed up together, indicating the break by the symbols of Fig. 56 or by the "limiting break line" (line 10 of the alphabet of lines), and giving the over-all dimension. Fig. 126, page 74, is an example of this principle applied to the drawing of a building.

Fastenings.

In every working drawing will occur the necessity of representing the methods of fastening parts together. Fastenings are either permanent, as rivets and nails, or removable, as bolts, screws, keys and pins. In drawing ordinary wooden structures the nails are not shown, unless there is some special reason for their location or number. Other fastenings, such as rivets, bolts, etc., are indicated by conventional symbols.

Bolts and Screws.

Bolts and screws are used for fastening parts together, for adjusting, and for transmitting power or motion. There are many different forms of bolts, and several different kinds of threads, for these different purposes. In drawing, we should know the conventional method of representing the ordinary types used.

The usual form of screw thread is the U. S. Standard, a V thread at 60 degrees, with the tip flattened and the root filled in. Among other forms are the sharp V, the square thread and the buttress thread, Fig. 57. When not otherwise specified, the U. S. Standard is always understood as required.

In ordinary drawing, threads are not drawn in actual form but are indicated conventionally. Fig. 58 shows several methods used. That shown in *A* is the simplest and best. It is not necessary to have the lines

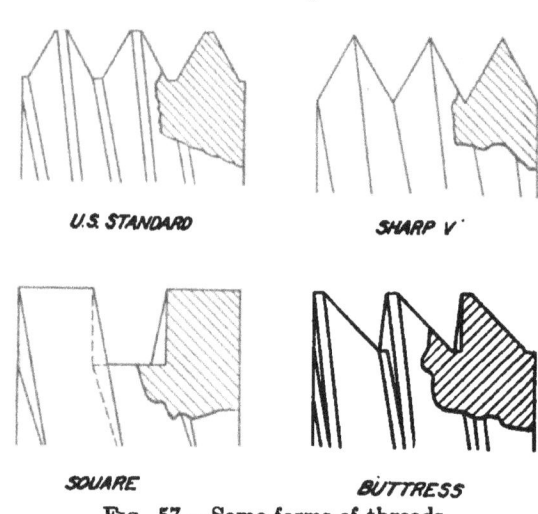

U.S. STANDARD   SHARP V

SQUARE   BUTTRESS

FIG. 57.—Some forms of threads.

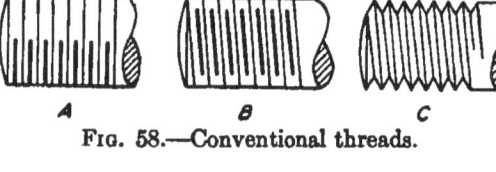

A   B   C

FIG. 58.—Conventional threads.

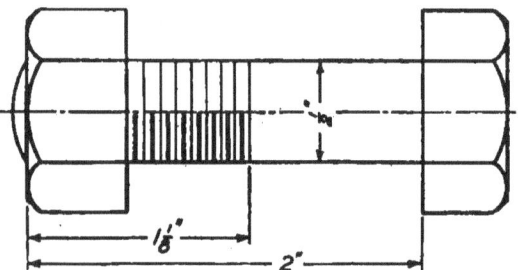

FIG. 59.—Dimensioned bolt.

spaced exactly to the pitch, or distance apart, of the actual threads, but to look well they should somewhat approximate it.

To Draw a Bolt.

There are adopted standards for standard hexagonal and square head bolts and nuts, hence in drawing, one view only is necessary,

and the only dimensions required are the diameter, the length (under the head) and the amount of thread, Fig. 59. In drawing a hex. head three faces are shown, and in a square head, one face. Figs. 60 and 61 are full size ¾″ bolts, given to show the proportions and method of drawing a hex. head and a square head bolt and nut, with the construction of the chamfer finish.

Fig. 62 illustrates a number of common forms of screw fastenings.

Pipes.

Pipe threads are cut on a taper, which is usually exaggerated slightly in drawing.

Pipe is designated by the nominal inside diameter, which differs slightly from the actual diameter, so that in drawing pipe we should know the outside diameter, and in figuring sizes needed, we should know the actual inside diameters or areas. The following table gives these sizes.

PIPE SIZES

Nominal inside diameter	Actual outside diameter	Actual inside diameter	Internal area
¼	.54	.36	.10
⅜	.675	.49	.19
½	.84	.62	.30
¾	1.05	.82	.53
1	1.315	1.05	.86
1¼	1.66	1.38	1.49
1½	1.9	1.61	2.03
2	2.375	2.06	3.35
2½	2.875	2.46	4.78
3	3.5	3.06	7.38

The method of drawing the usual fittings to different scales is shown in Fig. 63.

Developed Surfaces.

Sometimes it is necessary to draw a full sized pattern to which a piece of sheet metal may be cut, which when rolled or formed up will make a required piece. The operation of laying out the pattern from the drawing is called the *development* of the surface. In

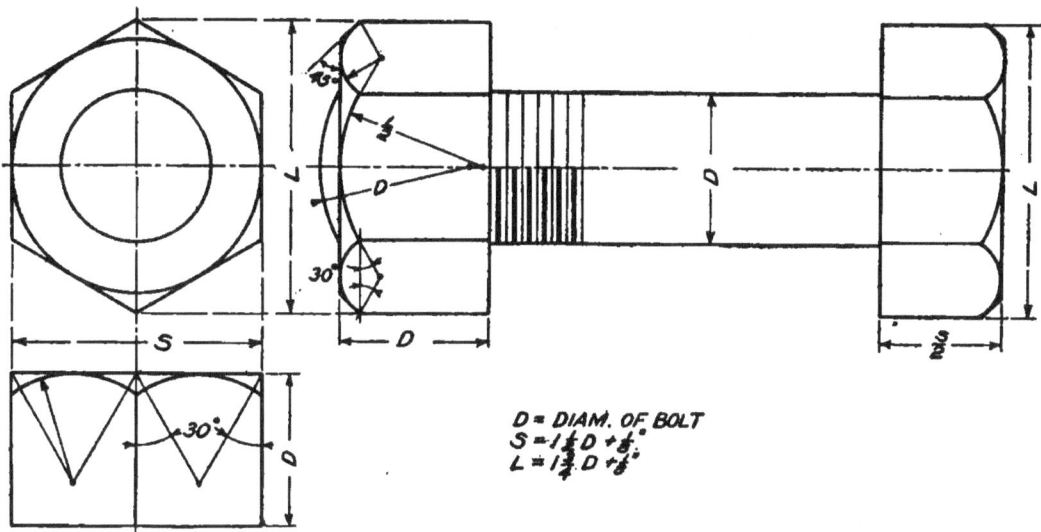

D = DIAM. OF BOLT
S = 1½ D + ⅛"
L = 1½ D + ⅛"

Fig. 60.—Construction of hexagonal head bolt.

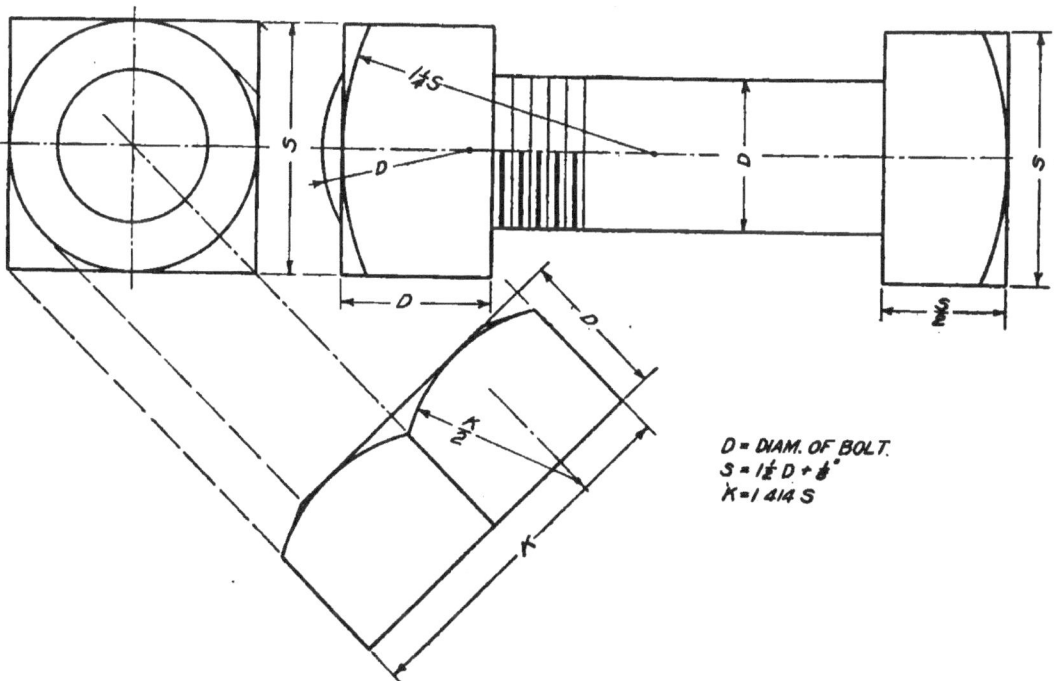

D = DIAM. OF BOLT.
S = 1½ D + ⅛"
K = 1.414 S

Fig. 61.—Construction of square head bolt.

our limited space we can only suggest the fundamentals of this branch of working drawings.

The pattern for a cylinder would evidently be a rectangle, whose width would be the length of the cylinder and whose length would be equal to the circumference, Fig. 64,

To develop a Cylinder. Fig. 66.

In rolling a cylinder out, the base will develop into a straight line, called the "stretchout line." Divide the base into a number of equal parts. Project these points as elements on the front view. Draw the stretchout line and step off the divisions of

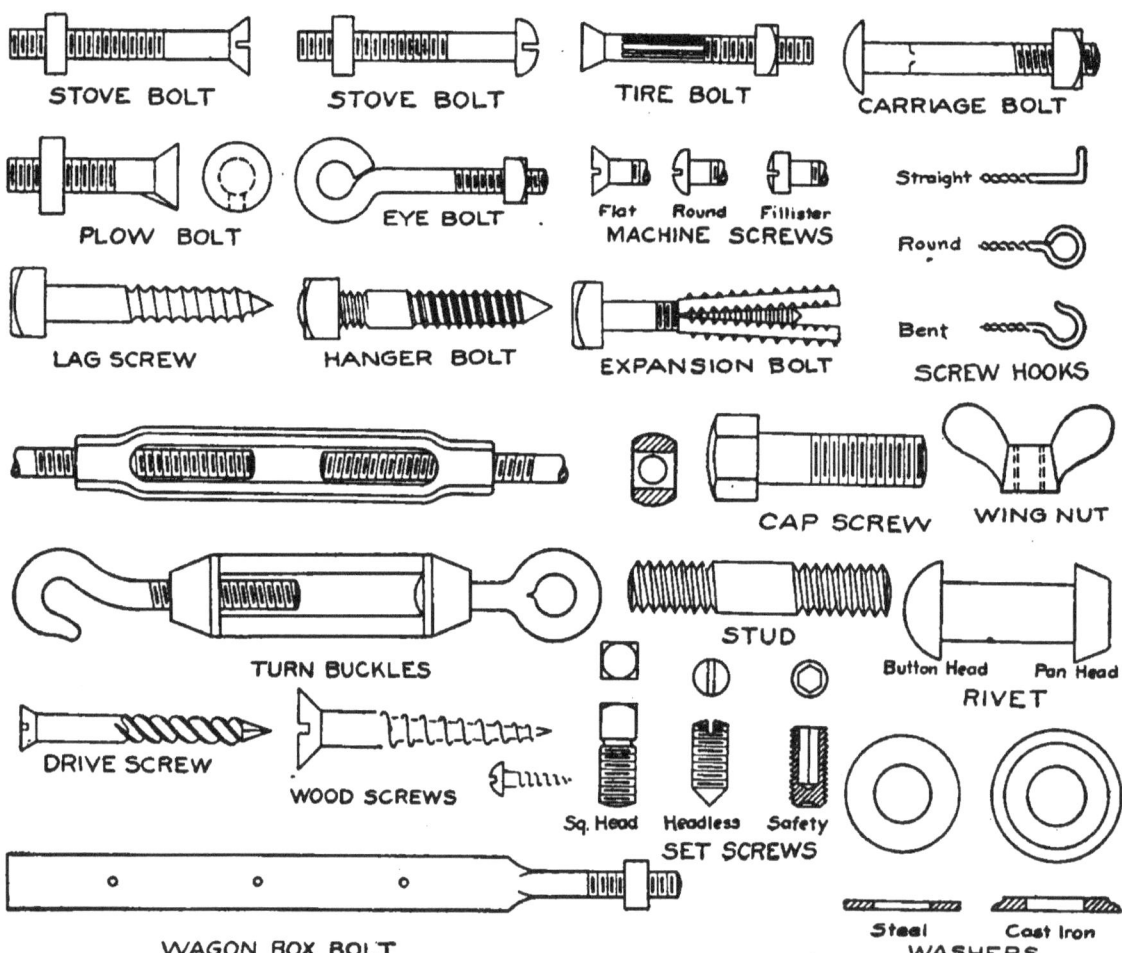

FIG. 62.—Various bolts and screws.

and the pattern for a cone would be a sector, with a radius equal to the slant height and an arc equal in length to the circumference of the base, Fig. 65.

In the development of any object we must first have its projections, and in practical applications must allow for seams and lap.

the base on it. Through these points draw the elements and project their lengths across in order from the front view. Connect these points by a smooth curve.

To develop a Truncated Cone. Fig. 67.

Divide the base as before. Draw the stretchout line with a radius equal to the

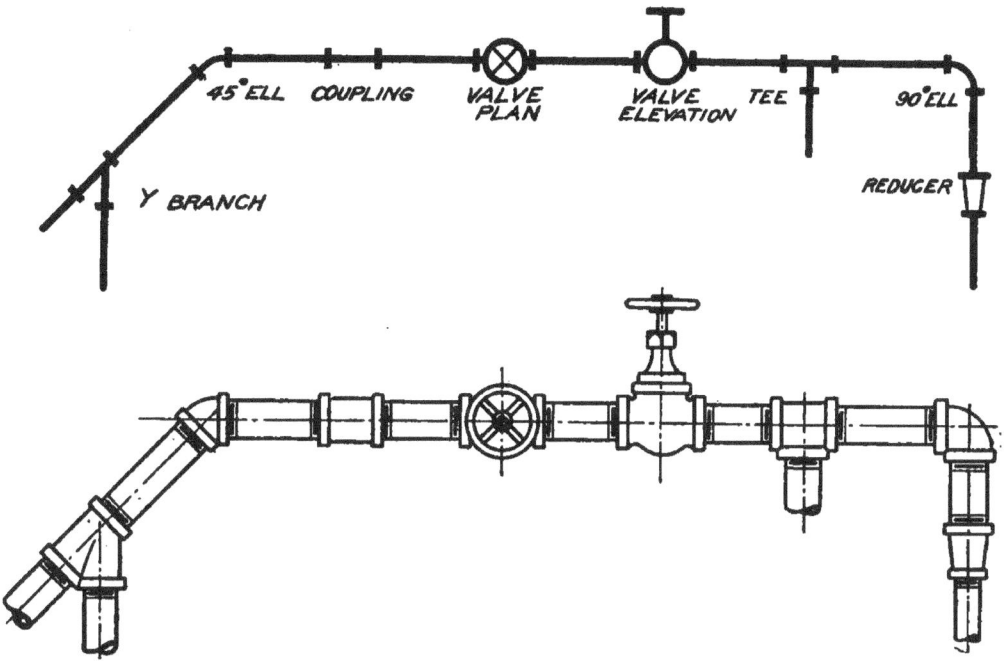

FIG. 63.—Pipe fittings, as drawn in small scale and large scale.

slant height of the completed cone, $O'A'$. Step off the divisions of the base on the stretchout line as shown and connect the two end points with the center O'. Then draw the arc for the small end with the radius $O'B'$.

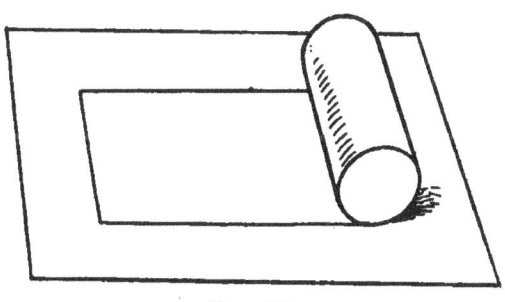

FIG. 64.

To develop a Prism.

Fig. 68 illustrates the development of the surface of a square prism with a sloping top, with the seam at the corner EA. The construction is evident from the figure.

To develop a Pyramid.

Fig. 69 shows the method of developing a truncated square pyramid with the seam

on EA. Complete the pyramid, finding the apex O. Since the edge EA is not shown in its true length on the front view, it must be revolved about the axis of the pyramid until it is parallel to the vertical plane, by swinging the top view of the line from OA to OA''

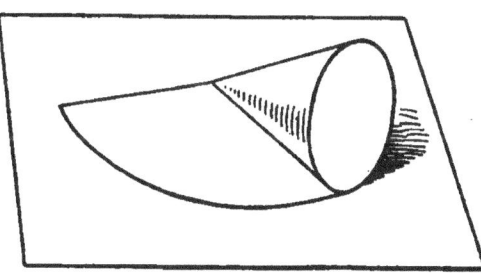

FIG. 65.

and projecting the point A'' down to A' Project E across to the line $O'A'$, then $E'A.'$ will be true length of the edge EA. (The pyramid may be imagined to be slipped inside of a cone having the same slant as the edges of the pyramid, and $O'A'$ will be the true length of an element of this cone.) With $O'A'$ as a radius, draw an arc as in the development of the cone and on this step off

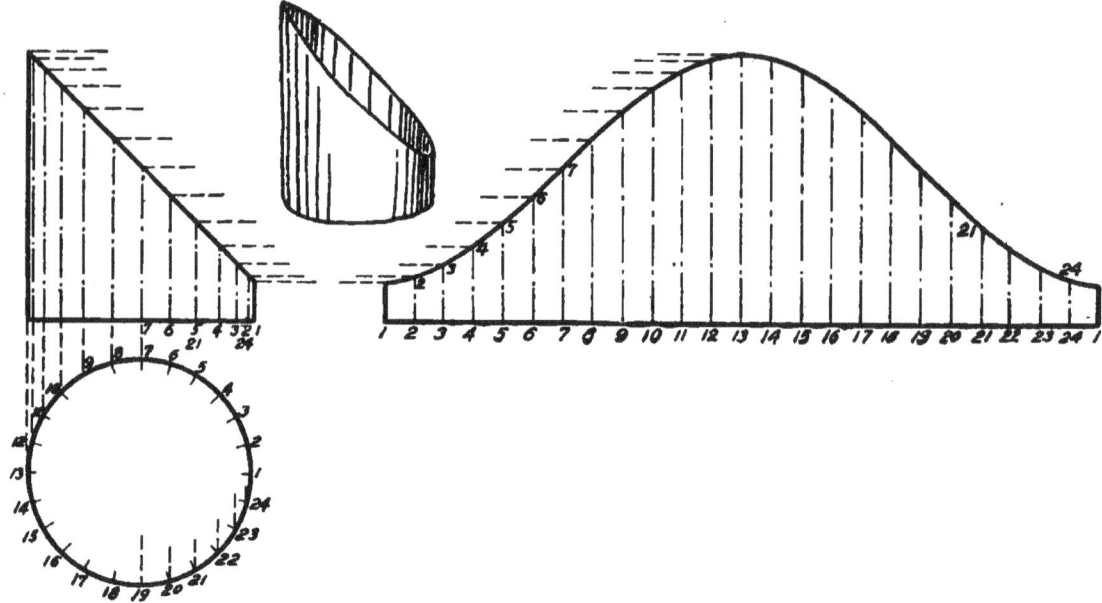

Fig. 66.—Development of a cylinder.

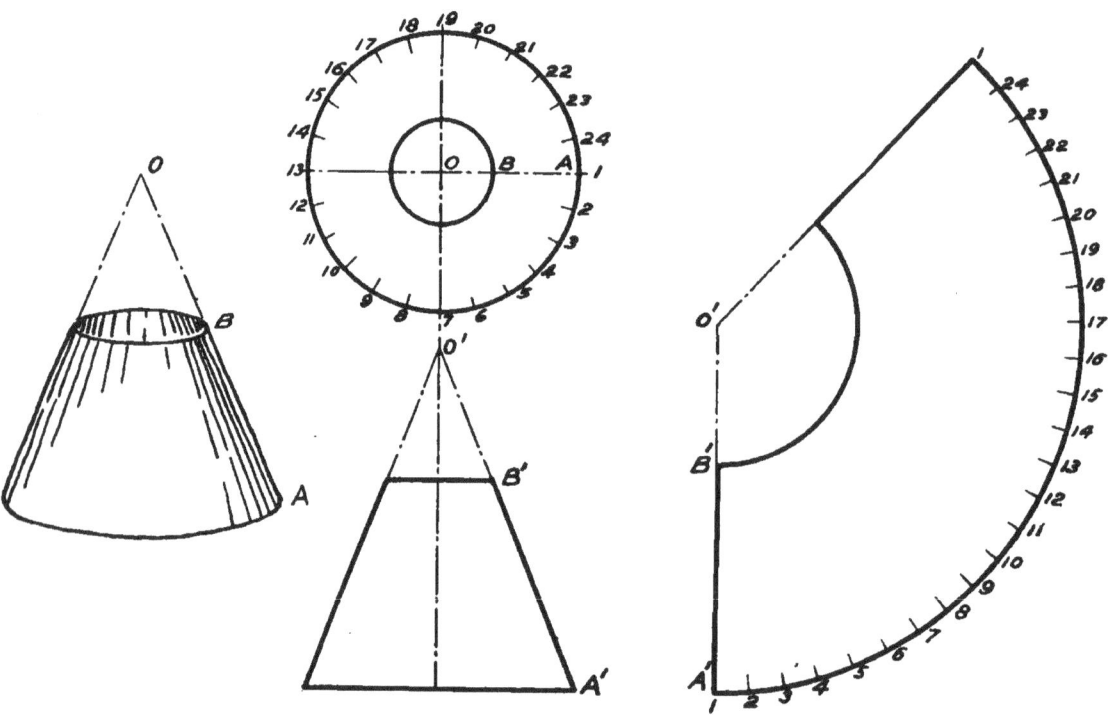

Fig. 67.—Development of a cone.

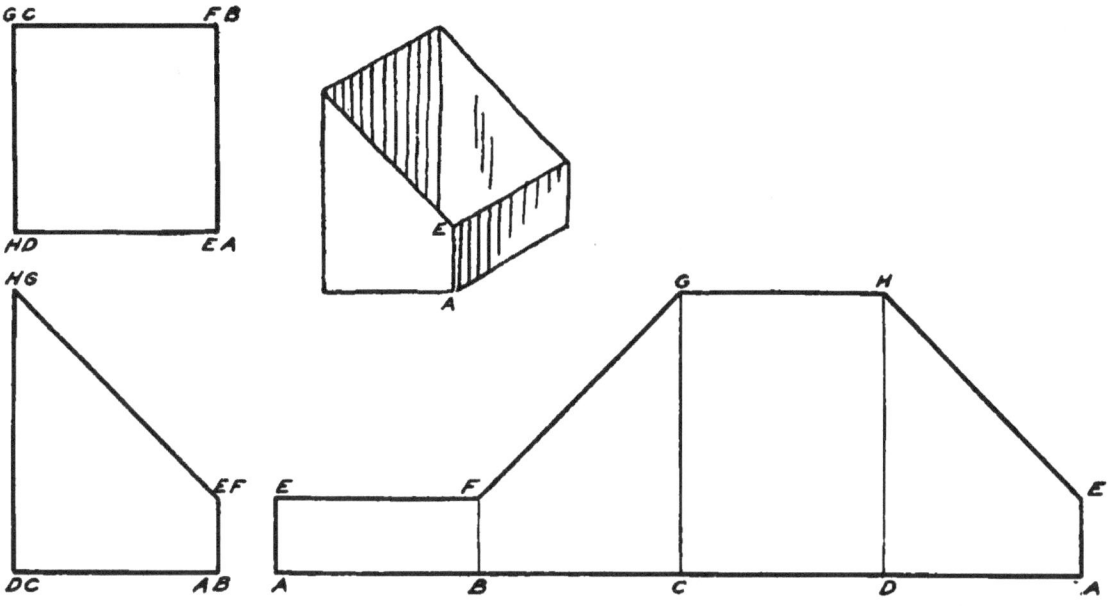

FIG. 68.—Development of a prism.

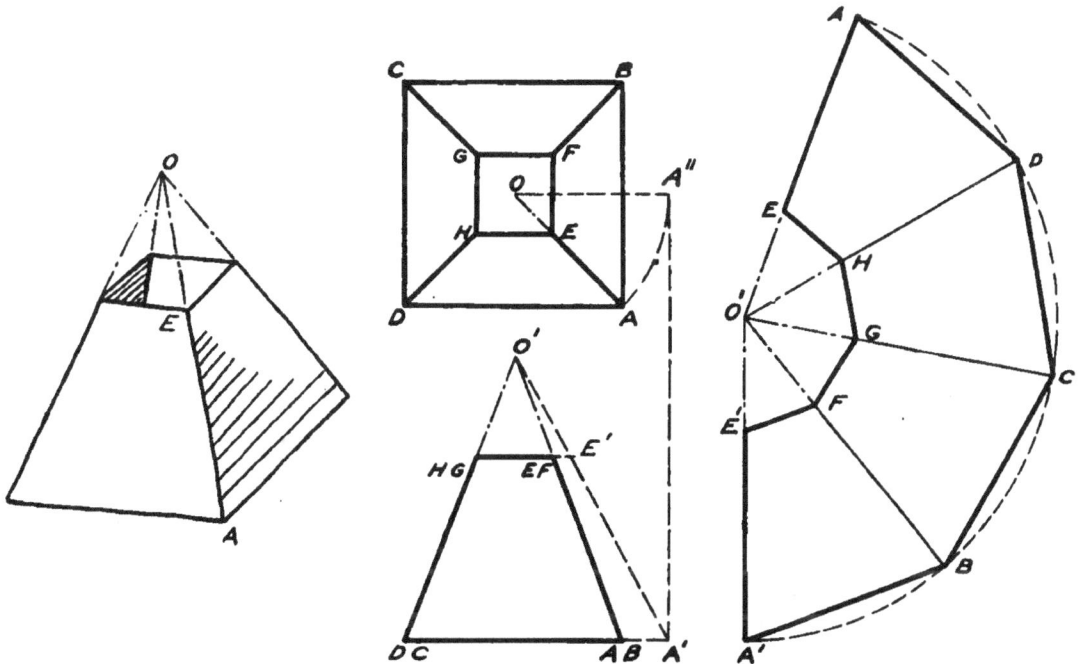

FIG. 69.—Development of a pyramid.

the four edges of the base of the pyramid *ABCD*. Connect the points *ABCDA* with *O* giving the folding edges. Find the development of the upper end by drawing an arc with the radius *O'E'*.

METHOD OF WORKING

In making a working drawing, after the scheming and inventing has been done in freehand sketches, the order of procedure in penciling should be about as follows: first, lay off the sheet with border line, and block out space for the title; second, plan the sheet, deciding upon the number and arrangement of views, always selecting as large a scale as possible; third, draw center lines for each view and lay off the principal dimensions; fourth, complete the projections; fifth, draw the dimension lines, and put in the dimensions; sixth, lay out the title; seventh, check the drawing carefully.

Order of Inking.

First, ink all circles, then circle arcs; second, ink the straight lines in the order—horizontal, vertical, inclined; third, ink center lines, extension and dimension lines; fourth, ink the dimensions; fifth, section line cut surfaces; sixth, ink notes, title and border line; seventh, check the tracing.

Sketching from the Object.

In our previous consideration of freehand sketching from pictorial views (page 9), we were concerned with the projections only. In connection with working drawings, it is often necessary to make a dimensioned sketch from the object itself, as for example, in the case of a broken piece of machinery.

The procedure would be somewhat in the order indicated. First, study the object and determine the necessary views; second, sketch center lines, observe the proportions and block in principal parts of the outline; third, finish the projections; fourth, draw all dimension lines and arrow-heads, before putting any figures in; fifth, measure the object and put the dimensions on the sketch; sixth, check for errors and omissions; seventh, date and sign the sketch.

Sometimes it requires considerable ingenuity to get accurate measurements. A two foot rule and calipers would be required for castings and small objects, and a steel tape for larger structures. A plumb line is often of service, and other devices will suggest themselves as the occasion demands.

The commonest fault in sketching is the overlooking of some important dimension whose omission is discovered as soon as the working drawing to scale is started.

PROBLEMS

In application of the principles of this chapter, selections from the following problems are to be made, and the complete working drawings with all necessary dimensions, notes, and title are to be drawn. Use a standard sized sheet (either 12″ × 18″ or 18″ × 24″), and follow the *method of working* as outlined on the previous page, selecting suitable scale, deciding upon the views necessary, and making preliminary sketch plan for the sheet. The drawings may be inked, or preferably traced on tracing cloth for blue printing. Some of the problems have complete data and dimensions given, others are intended to be designed by the student.

Be particularly careful to follow the rules for dimensioning.

Problems

1. Make the working drawing of a box for fence repair kit, from the sketch, Fig. 70. (Of course the words "hammer," etc., will not appear on the drawing.)

2. Working drawing of germinating or corn testing box, from the sketch, Fig. 71. The bottom is rabbeted so as to be water tight, and the saw cuts are to hold cords for making divisions.

3. Working drawing of heavy clevis, Fig. 72. Specify wrought iron or mild steel.

4. Working drawing of sheep feeding rack, Fig.

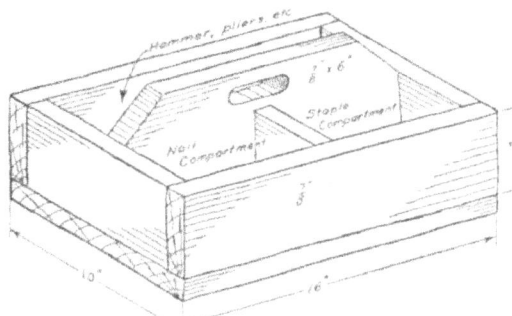

FIG. 70.—Tool box.

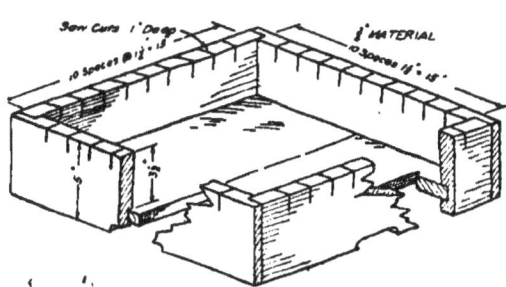

FIG. 71.—Corn tester.

73. Decide length wanted and space sills not more than six feet apart.

5. Working drawing of sheep feeding rack using alternate form shown in Fig. 74. This has larger capacity than Fig. 73, and the flaring boards prevent dirt from getting into the wool on the sheep's neck.

6. Working drawing of milk stool, Fig. 75.

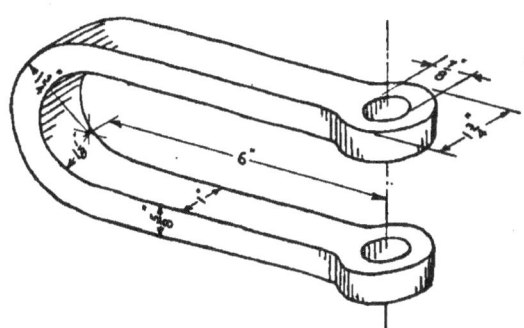

FIG. 72.—Clevis.

7. Working drawing of sack holder, Fig. 76. A mortised wood brace may be substituted for the wrought iron brace. The hooks may be made of finishing nails.

8. Working drawing of gang mold for test pieces, Fig. 77. This mold is for making standard $1\frac{1}{2}'' \times 1\frac{1}{2}'' \times 6''$ test specimens used in the rough determi-

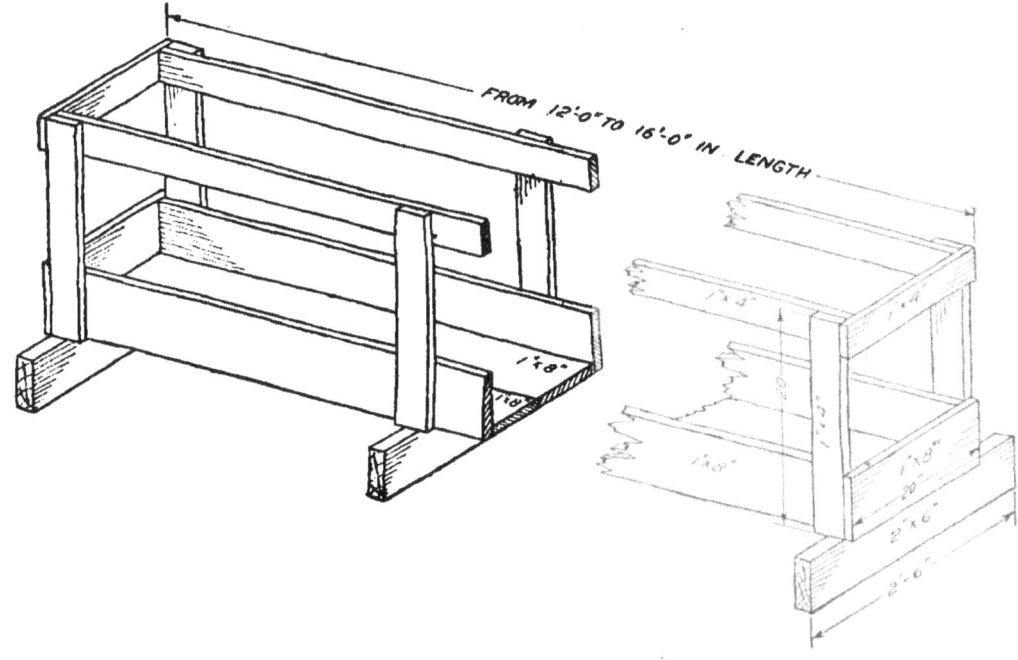

FIG. 73.—Sheep rack.

nation of the breaking strength of cement. Detail such parts as are necessary for showing construction.

9. Working drawing of forms for concrete hog trough, Fig. 78 (the trough is cast inverted).

10. Working drawing of plank road drag, Fig. 79. Use conventional symbol in showing a ⅜″ chain 7 ft. long, attached to hooks.

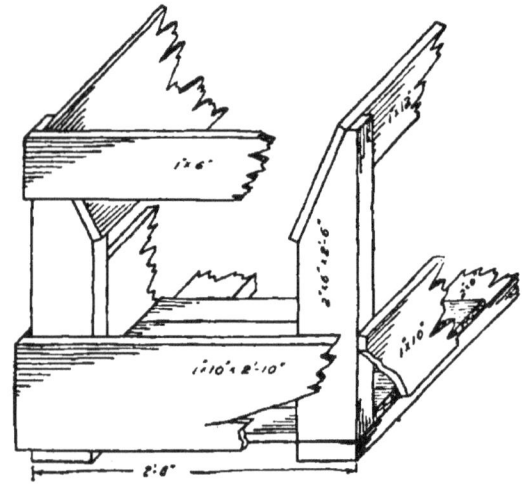

Fig. 74.—Sheep rack.

11. Working drawing of split log drag, using same dimensions as in Problem 10.

12. Working drawing of low-down silo rack, Fig. 80. The wrought iron end plates are ½″ stock with 1¼″ hole and fastened with ⅝″ bolts. The rack is swung under the axles of a wagon from which

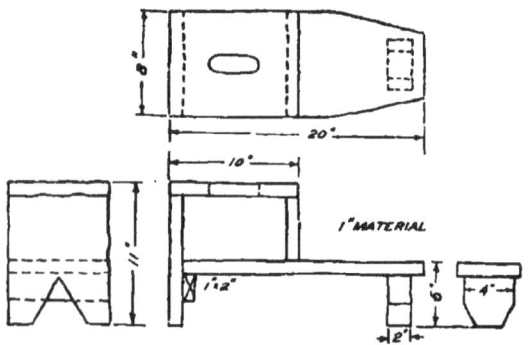

Fig. 75.—Milk stool.

the bed and reach have been removed. The bed pieces are hung to the rear axle by clip irons.

13. Working drawing of stone boat, Fig. 81, using auxiliary projection for nose.

14. Working drawing of boot jack, Fig. 82 (side view and auxiliary projection are the only views needed).

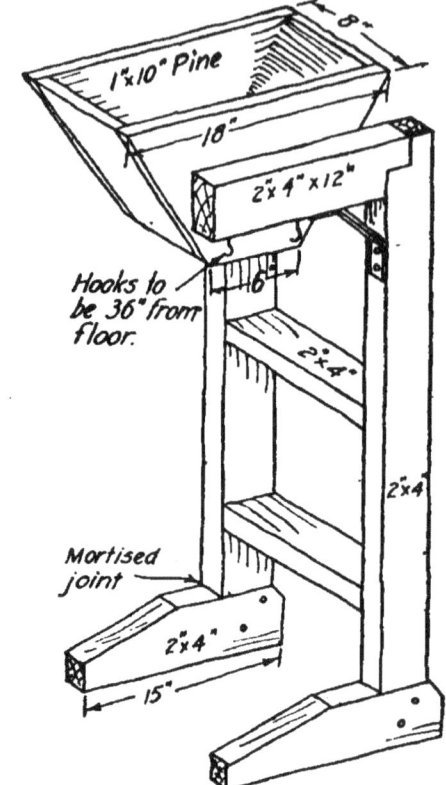

Fig. 76.—Sack holder.

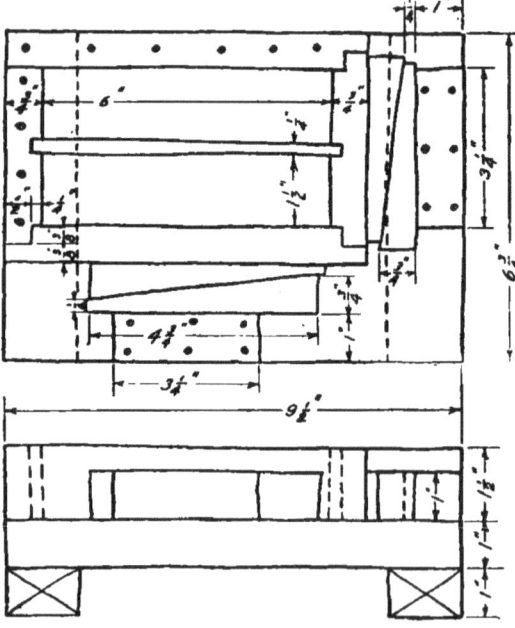

MATERIAL:
WHITE PINE. APPLY A MIXTURE OF
EQUAL PARTS BOILED LINSEED OIL
AND KEROSENE

Fig. 77.—Gang mold.

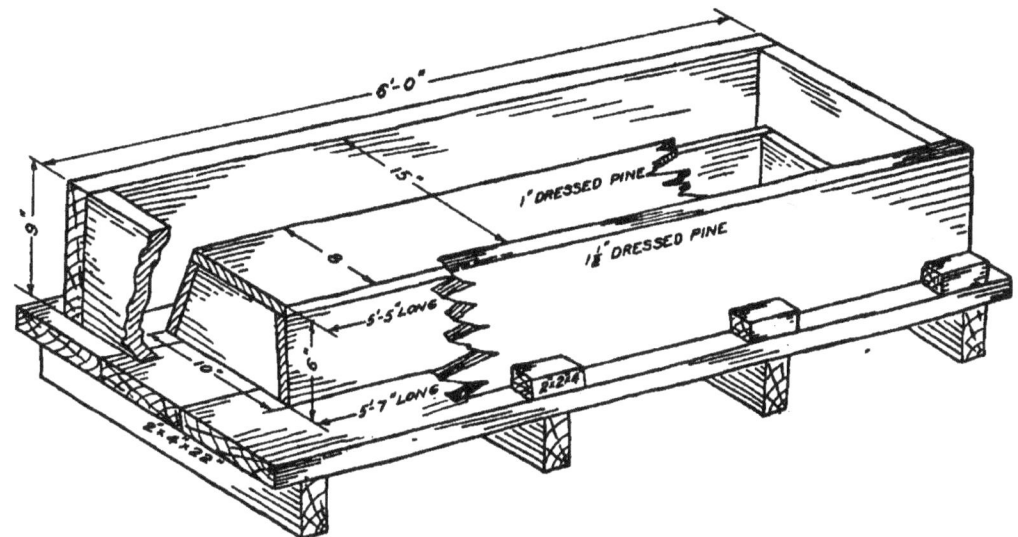

Fig. 78.—Form for concrete trough.

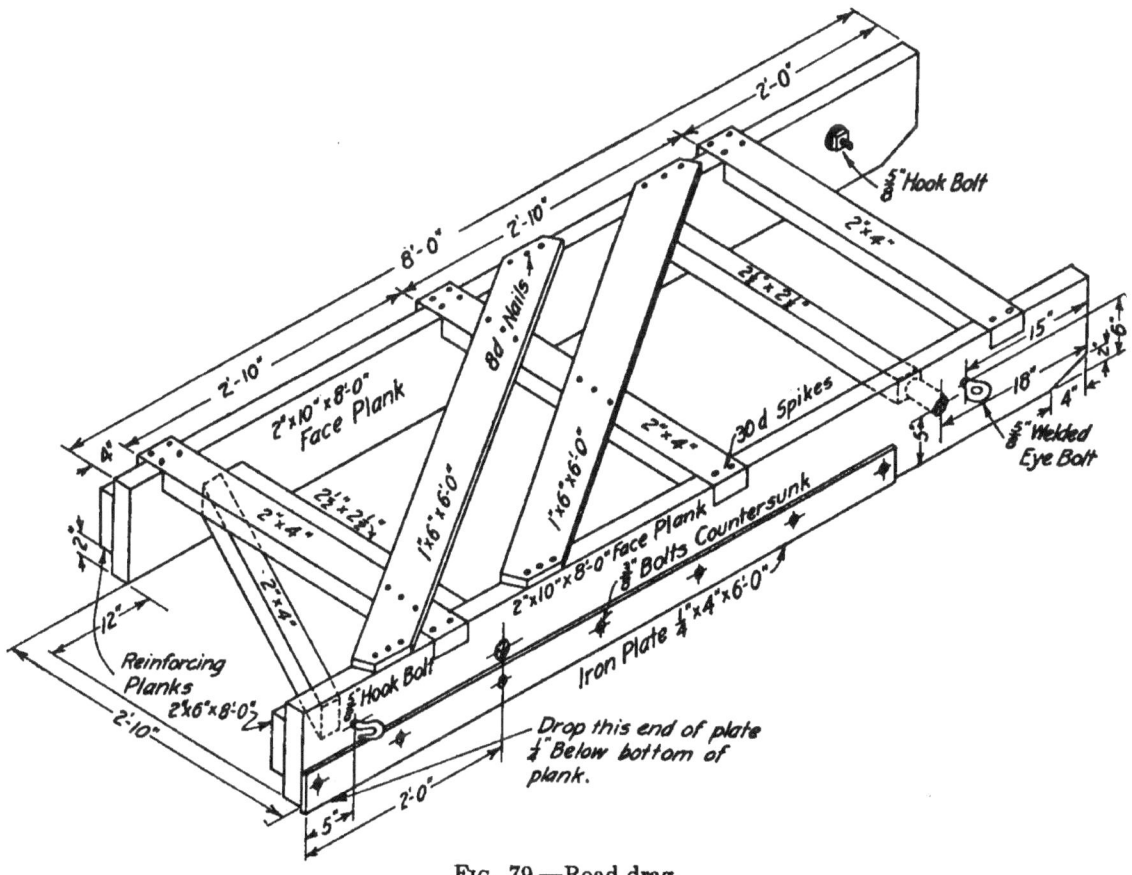

Fig. 79.—Road drag.

15. Working drawing of fly wheel, Fig. 83 (front view and section).

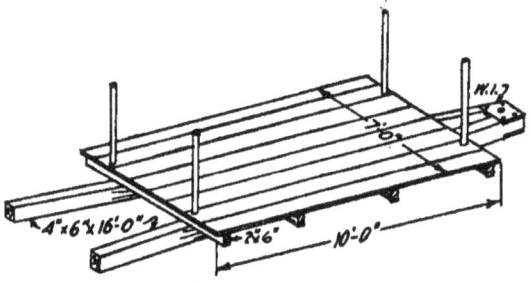

Fig. 80.—Silo rack.

16. Working drawing of flanged roller, Fig. 84 (front view and section).

17. Working drawing of home made muffler for

19. Working drawing of ice box, Fig. 87. This design is called the Rochester cold box, and is easily built and very efficient.

20. Working drawing of farm gate, Fig. 88. Length may be varied if desired.

21. Working drawing of farm gate of your own design. · Suggestions of different forms may be had from Fig. 133, page 81.

22. Working drawing of grindstone and frame, Fig. 89, making assembly and detail drawings.

23. Working drawing of sheave, Fig. 90. Show right half of side view in section.

24. Working drawing of cattle breeding crate, Fig. 91, including bill of material.

25. Working drawing of fly trap, Fig. 92. The bait holders d are can lids.

26. Develop patterns for three piece elbow, Fig. 93.

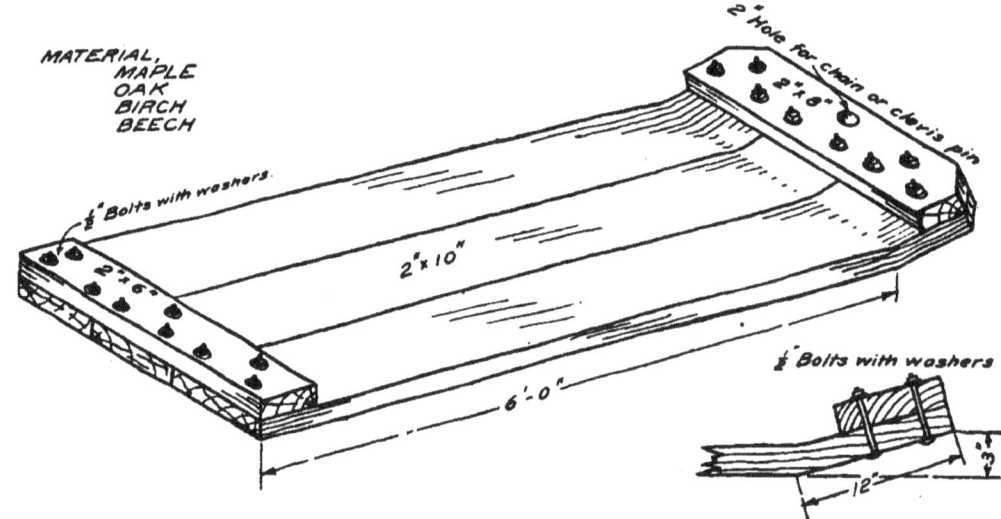

Fig. 81.—Stone boat.

gas engine, Fig. 85 (for dimension of pipe, see table of pipe sizes).

18. Working drawing of road, or community

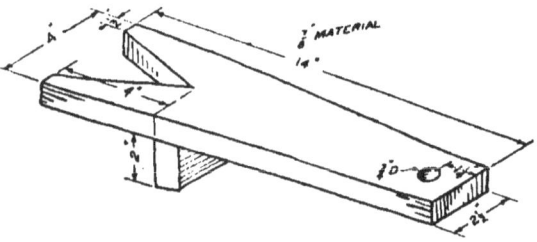

Fig. 82.—Boot jack.

bulletin board, Fig. 86. This drawing may be fully completed by adding in upright Gothic letters, the lettering to be painted on the board.

27. Develop pattern for conical reflector, Fig. 94.

28. Develop pattern for hexagonal lamp shade, Fig. 94.

29. Develop pattern for funnel, Fig. 95.

30. Develop patterns for sheet metal hopper, Fig. 95.

Models may be made of problems 26 to 30 by cutting the patterns out of paper and rolling or forming them up and pasting. If this is done, allowance must be made for lap.

31. Make a working drawing of a hay rack. The two bed or body pieces are to be 2″ × 8″ hard pine sixteen feet long set on edge. Across these place on edge five 2″ × 6″ pieces seven feet long for narrow rack, or eight feet for wide rack. Floor the whole with 1″ material. For securing the cross pieces to the bed, use clip irons as shown in Fig. 20 (M and

N) page 12. The bed pieces should be placed so as to overhang the front bolster eighteen inches, and should be twenty-four inches apart in front to allow a wide turning radius for the front wheels, and flare back to full width, forty-two inches, at the rear bolster. The front should be provided with a

forty-two inches long at each end of every cross-piece, securing them with bolts or irons. To these fasten three fence boards lengthwise and spaced evenly up and down. On the end standards nail similar boards and secure the corners with large hasps and staples.

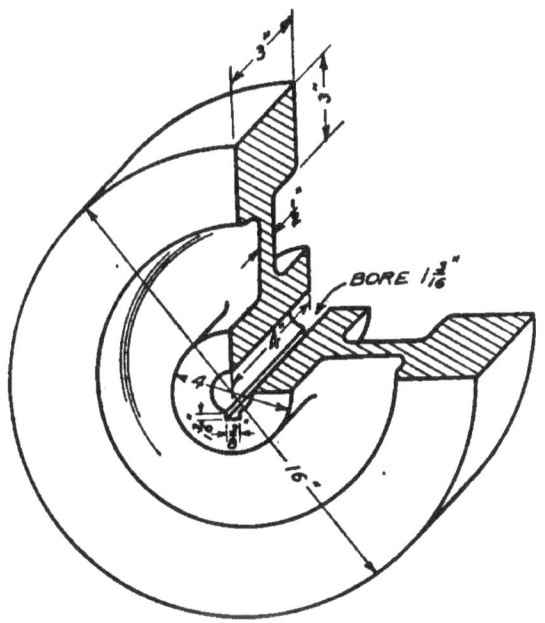

FIG. 83.—Flywheel.

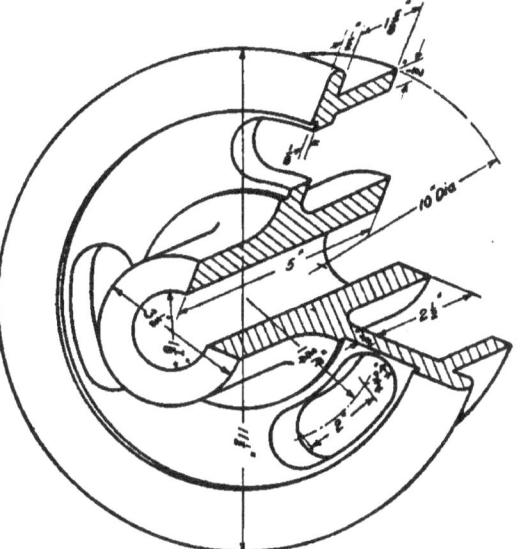

FIG. 84.—Flanged roller.

false bolster to prevent shifting from side to side. A standard, made of 2″ × 4″ pieces six feet long will be required.

32. Design a basket rack. A simple basket rack may be made as follows. On the hay rack described in Prob. 31, erect 2″ × 4″ pieces thirty-eight to

33. Make a working drawing of a gate post mold for 12″ × 12″ × 8′-0″ concrete post. The post to be reinforced by four ¾″ rods wired at each foot of length.

34. Working drawing of gang mold for fence posts seven feet long with 4 × 4 top and 4 × 6 bottom, posts to be reinforced with four ¼″ rods. Rods not to be closer than 1″ to outside.

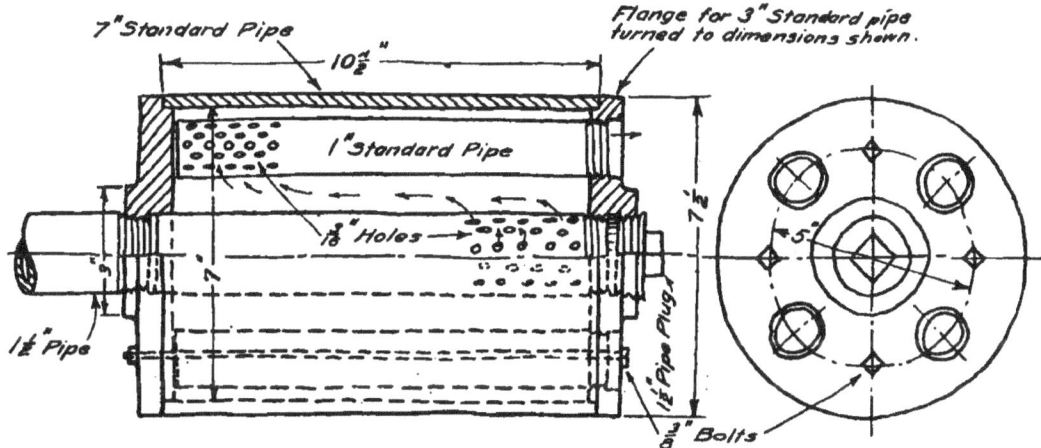

FIG. 85.—Muffler.

35. Design a work bench 34″ high, using the following material—

Top.......... 2 pcs. 1 × 14 × 8′- 0″
Aprons....... 2 pcs. 1 × 12 × 8′- 0″
Cross Boards.. 4 pcs. 1 × 12 × 2′- 2″
Legs.......... 4 pcs. 2 × 4 × 2′- 9″
Cross Ties..... 2 pcs. 1 × 4 × 2′- 2″
Wood vise screw 8″ from top.

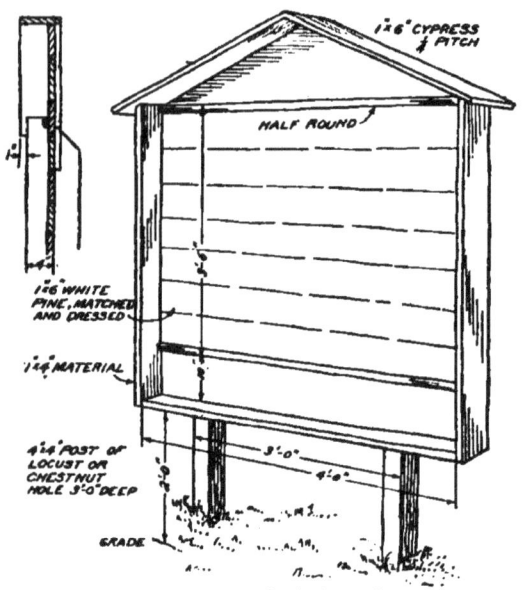

FIG. 86.—Bulletin board.

farm where the water is piped from a well to a sink in the kitchen; from the well to a tank in the barn yard, and off this line three outlets, one to horse stable, one to hog house and one to dairy house. The house supply pipe is 1″, that to tank 2″; all others ½″ pipe. Note on drawing and make list of all necessary fittings.

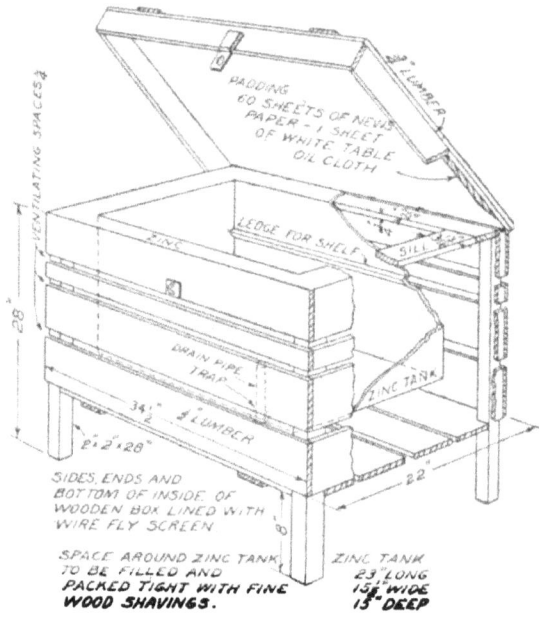

FIG. 87.—Ice box.

36. Design a pipe railing for the approach to a bank barn—use 1½″ pipe and the following fittings: cross, tee, 90° ell, flange.

37. Design a layout of the water piping for a

38. Design a cold frame, using 34″ × 52″ sash.

39. Design a suitable wood box for firewood. Box to be of neat design to match woodwork in kitchen. A sloping, hinged lid should be provided.

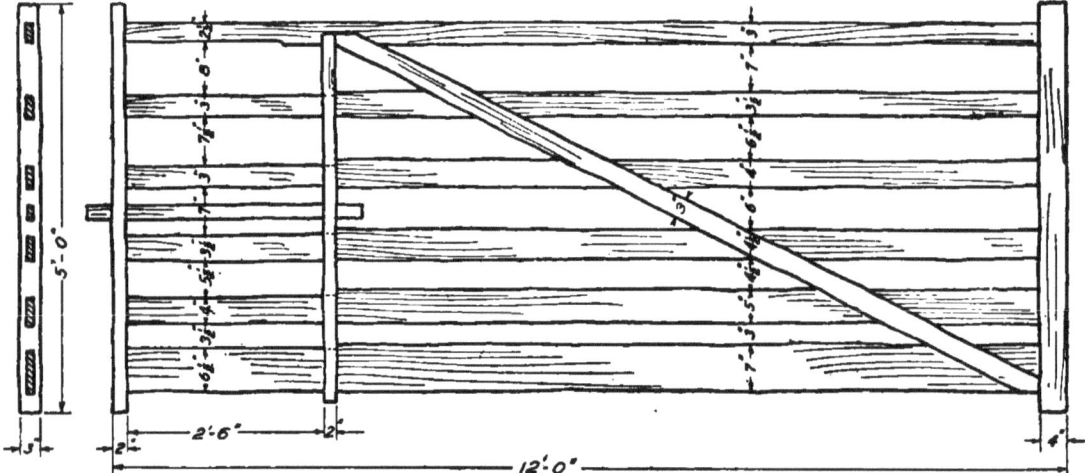

FIG. 88.—Farm gate.

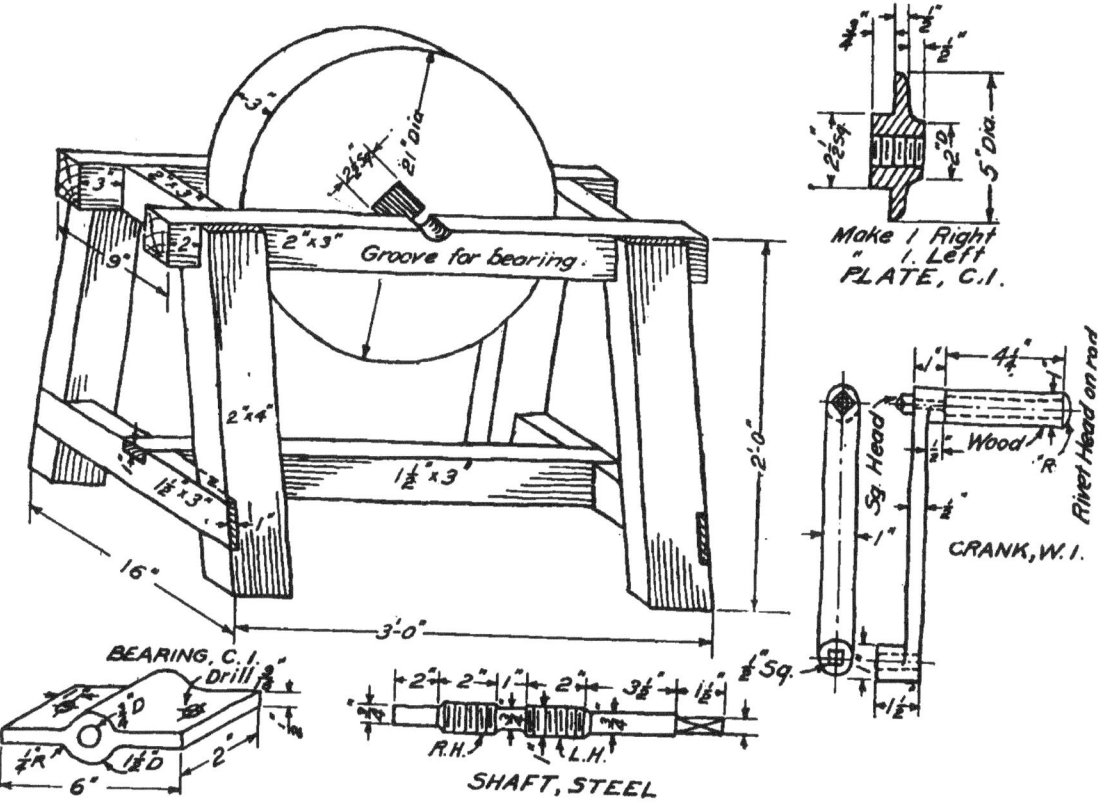

FIG. 89.—Grindstone frame details.

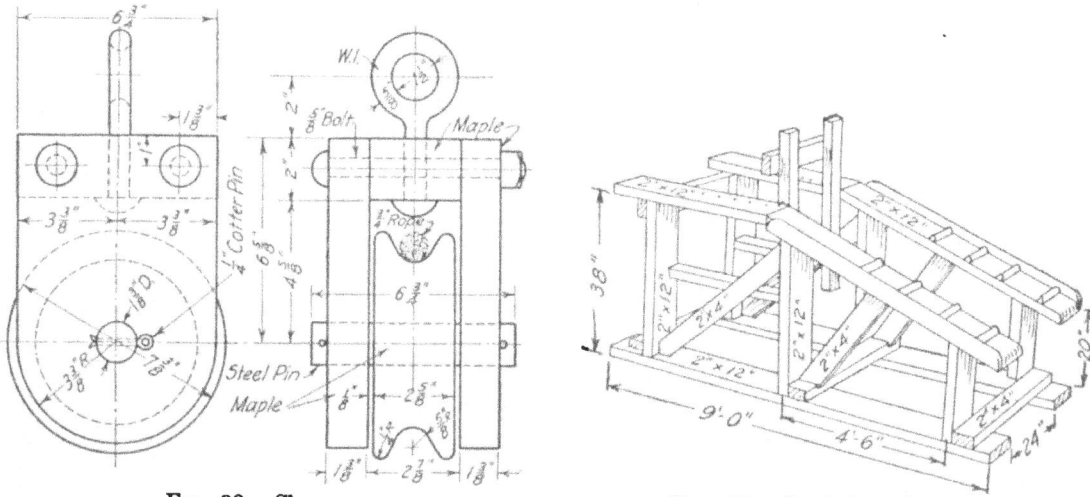

FIG. 90.—Sheave.   FIG. 91.—Cattle breeding crate.

40. Design and draw a trap nest. The Connecticut trap nest is illustrated in Fig. 2, page 2. Nest should be at least 12″ square in the clear.

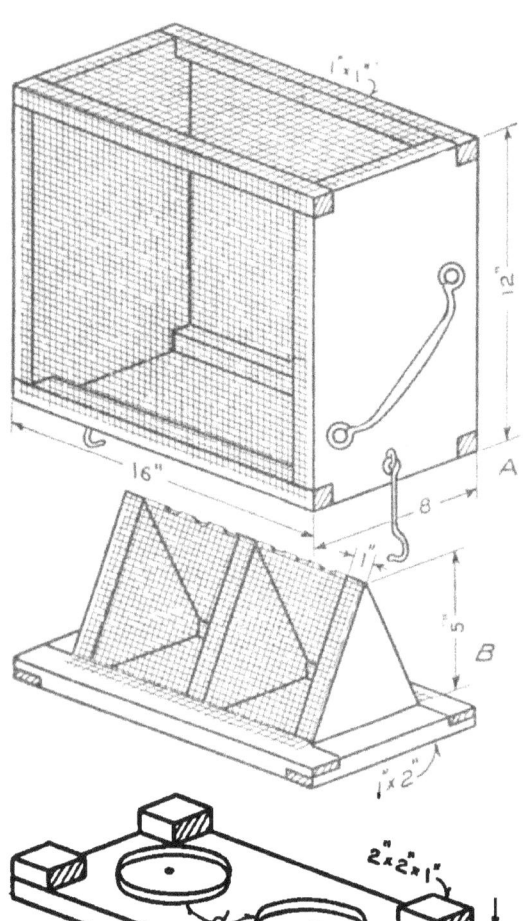

FIG. 92.—Homemade fly trap.

chest. The two till chest shown in Fig. 180, page 110, may serve for suggestion.

Additional working drawing problems may be

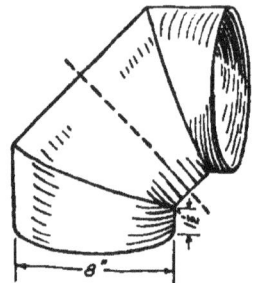

FIG. 93.—Three-piece elbow.

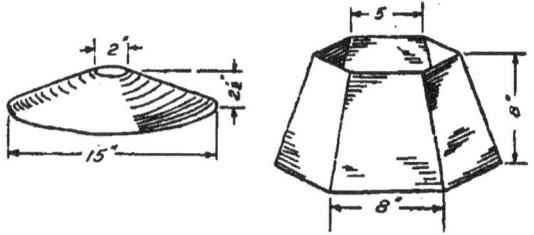

FIG. 94.—Reflector and lamp shade.

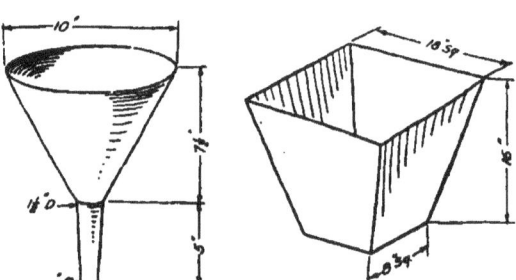

FIG. 95.—Funnel and hopper.

41. Design a wash bench 16″ high, 3 ft. long, with wringer board in the middle.

42. Design and make working drawing of a tool

made from Figs. 160 to 181 in Chapter VI, and may be suggested by other illustrations throughout the book.

CHAPTER IV

FARM STRUCTURES

Under the head of farm structures would properly be included any structural work built for confining, enclosing or covering animals, machinery, or materials, and of course including dwellings. Thus fences, paddocks, pens, gates, etc., would be called confining structures, while enclosing and covering structures would include houses, barns, and sheds, together with any building work which would improve them or make them more efficient; for example, a manure pit and a silo would add to the efficiency of a dairy barn.

In this chapter we are to consider the methods of designing and drawing these structures.

Designing is simply suiting form to function. Thus in the design of any structure we must first consider the reasons and uses for it, following mentally in detail all through the operations which will be performed in it, and making it to such form that it will serve its full purpose to the best advantage possible, and be practical and economical both in construction and operation. After the general design has been thought out, the drawing of it is the application of the principles of the language which we have been studying. Training in this systematic process of reasoning, and in the ability to think on paper is of particular value to the farmer. It gives him the power to plan for improvements, to estimate costs, and to read intelligently plans presented to him.

The actual execution of these structural drawings is comparatively simple, as they are made up principally of straight lines. After one is familiar with the instruments and the theory of working drawings, all that is needed in making drawings of structures is a facility in the use of the scale and a knowledge of the conventional symbols used in representing different materials and different parts. The drawing of larger structures must be done to such small scale that the details of these parts can not be shown but are only indicated by standard signs. Their construction is shown by detail drawings to larger scale, which are usually grouped on separate sheets.

Symbols.

On page 30 the use of conventional symbols was explained, and examples used in working drawings were illustrated. In structural drawing we need to know the conventional symbols for doors, windows, etc., as well as the symbols for sections of different materials. Fig. 96 gives a number of these symbols, whose application will be needed in reading the drawings following, and in drawing the problems given.

Plans.

As mentioned in Chapter II the terms *plan* and *front, side* or *end elevation* are used instead of top view, front view and side view. The plan of a building is really a horizontal section as if the building were cut through at the height of the windows and the top lifted off. This cutting plane may be

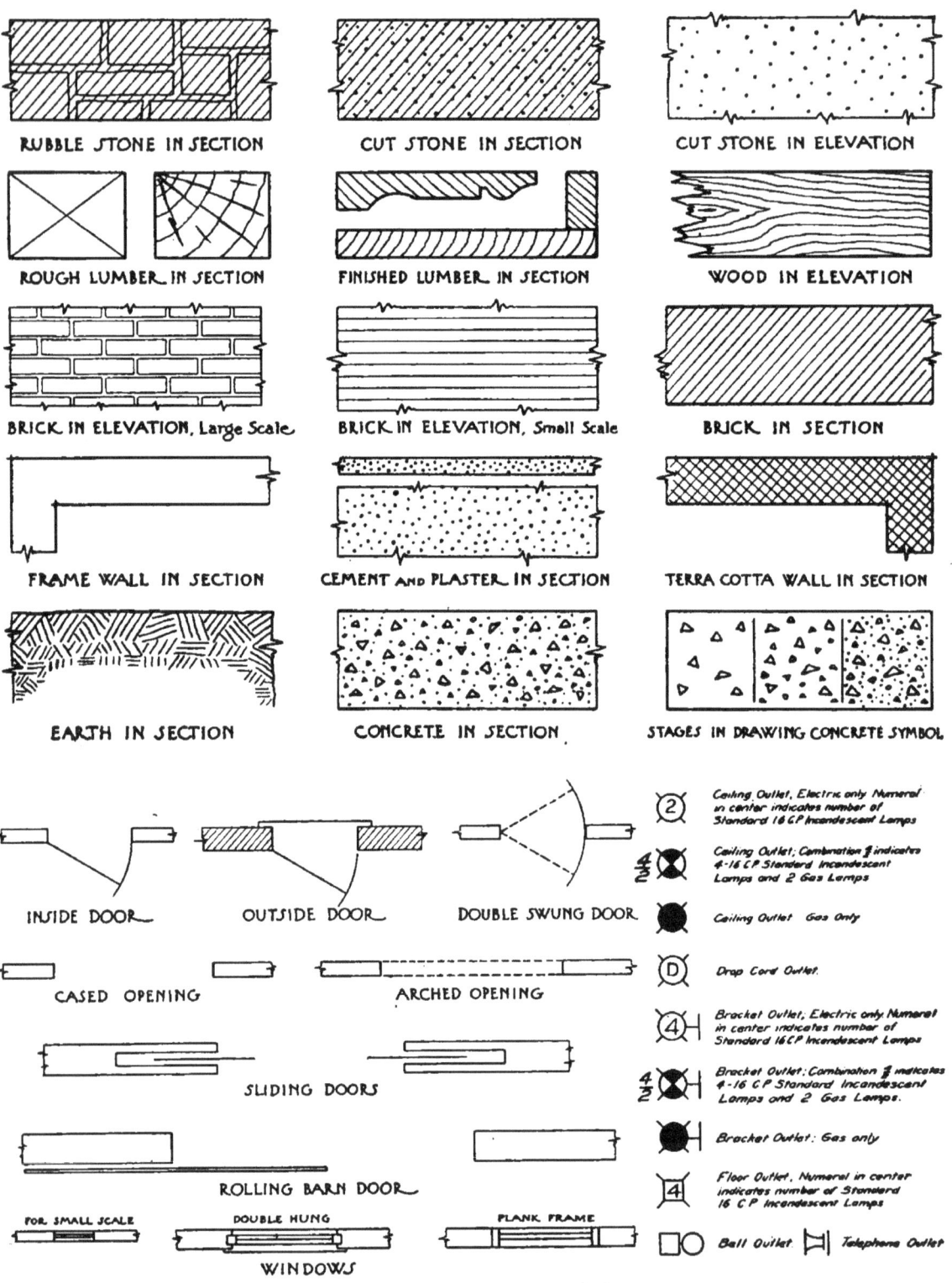

Fig. 96.—Architectural symbols.

assumed to vary in height at different points in order to show the parts of the plan to the best advantage. This is illustrated in Fig. 97 where it passes through the fireplace, then through a high window.

Plans of houses and larger buildings are usually drawn to the scale of $\frac{1}{4}'' = 1$ ft. (sometimes written $1'' = 4$ ft.), and generally one view only is placed on a sheet. Drawings of smaller structures may be made to $\frac{1}{2}''$ or even $1''$ scale, and all the views shown on one sheet.

"wall section" should always first be laid off at one side, in order to determine the heights of windows from the floor, heights of floors, cornice lines, etc., and these points projected across to the elevation. This wall section is often left on the finished drawing, to show heights, and cornice construction.

Sections.

Vertical sections are often necessary to show interior construction, such as framing,

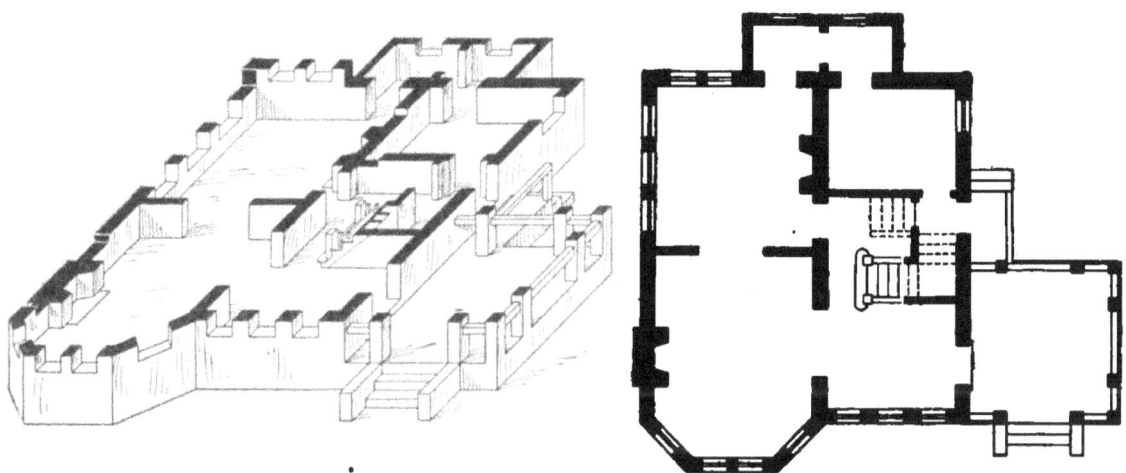

Fig. 97.—Pictorial illustration of a "plan."

In planning a new building the thinking and scheming should be done with the aid of rough freehand sketches, using single lines for the walls, before a scale drawing is started. Cross section paper is often used in sketching, to aid in getting correct proportions. No drawing should be finished in ink until all the sheets are complete in pencil, as often a change on one view will necessitate changing several others.

Elevations.

An elevation is a front or side view of a building, showing the exterior appearance, and is of use in indicating the location of windows and doors, glass sizes, heights of floors, etc. In drawing an elevation, a

stairways, trusses and other details. Numerous examples of vertical sections are given in this chapter. The terms "longitudinal section" and "transverse" or "cross section" are sometimes used to indicate that the section is taken lengthwise or crosswise.

Dimensions.

As in other working drawings, dimensions and notes are most important. The rules for dimensioning given on page 29 should be applied, and the needs of the builder always kept in mind. Thus dimensions should always be given to and from accessible points so that they may be laid off accurately and without computation.

Dimensions for building work should be given in even feet and inches, avoiding fractions of inches as far as possible, except as they may be necessary in details.

Commercial and stock sizes should be used as far as possible on account of economy. For example, a piece of glass 7" × 10" costs more than one 8" × 10" because the latter is a stock size while the former has to be cut to order. Some tables of commercial sizes for reference will be found on page 112.

Plan of the Chapter.

In the plan of this chapter, a short discussion of different building materials and their uses, and of general methods of preparing drawings, will be followed by a discussion and illustration of each of several typical farm structures, representing modern practice in design and construction; the first one giving complete plans, specifications, bill of material, and estimate, as an example to be followed. The application of the principles explained is to be made in the problems given at the end of the chapter. Many variations of the problems as stated, will suggest themselves to the student, and it is hoped that the varied selection of examples and the description of requirements and present practice will enable him to adapt the principles and make dependable working drawings of any ordinary structure needed.

Wood Construction.

By far the greatest number of farm structures are at present built of wood. The wood used is largely from local timber, but its gradual diminution in many sections, is leading to the use of southern and western lumber.

Wood is classified in a general way commercially, as hard wood and soft wood, hard wood being from deciduous trees, and soft wood from cone-bearing trees. The designer will select, and specify on the drawing, the available wood best suited to the particular purpose. If used for carrying load or stress, the sizes needed in the wood selected must be figured for safe load. The table of comparative strength of timbers and the loads they will carry given on page 114, will be of use in these computations.

A list of the kinds of wood used for different purposes, somewhat in the order of their desirability and availability, is given on page 113.

In making a drawing for a framed structure it is necessary to show clearly the method of construction. The sizes and lengths of timbers must be given in order that the quantities may be taken off, and to prevent the cutting up of special timbers in places where shorter lengths have been designed. The kinds of joints intended must be shown, and if necessary supplemented by details.

In large construction, such as barn framing, there are two general systems, the braced, pin-joint frame, made of heavy timbers, and the plank frame, made up of two inch planking, either in the form of the "plank truss" or the "balloon frame."

In the structures which follow, these different methods of framing are illustrated. The braced frame is shown in the horse barn, Fig. 116, the plank truss in Fig. 114, and the "balloon frame" in Figs. 117 and 127.

The timber frame is the older type and is very substantial and reliable. It is used in localities where large squared timber of good quality and sufficient length is readily available. The plank frame is adapted to sections where timber is purchased or must be hauled long distances.

Regarding the advantages and disadvantages of the two systems, workmen through the country are more familiar with timber framing. Often they are not at all familiar with plank frame construction. Instances of collapse are numerous, due to faulty workmanship and lack of information regarding plank frame requirements.

The advantages of the plank frame are the time saved in cutting out and erection. Also fewer men are required for raising. While it is comparatively simple to obtain

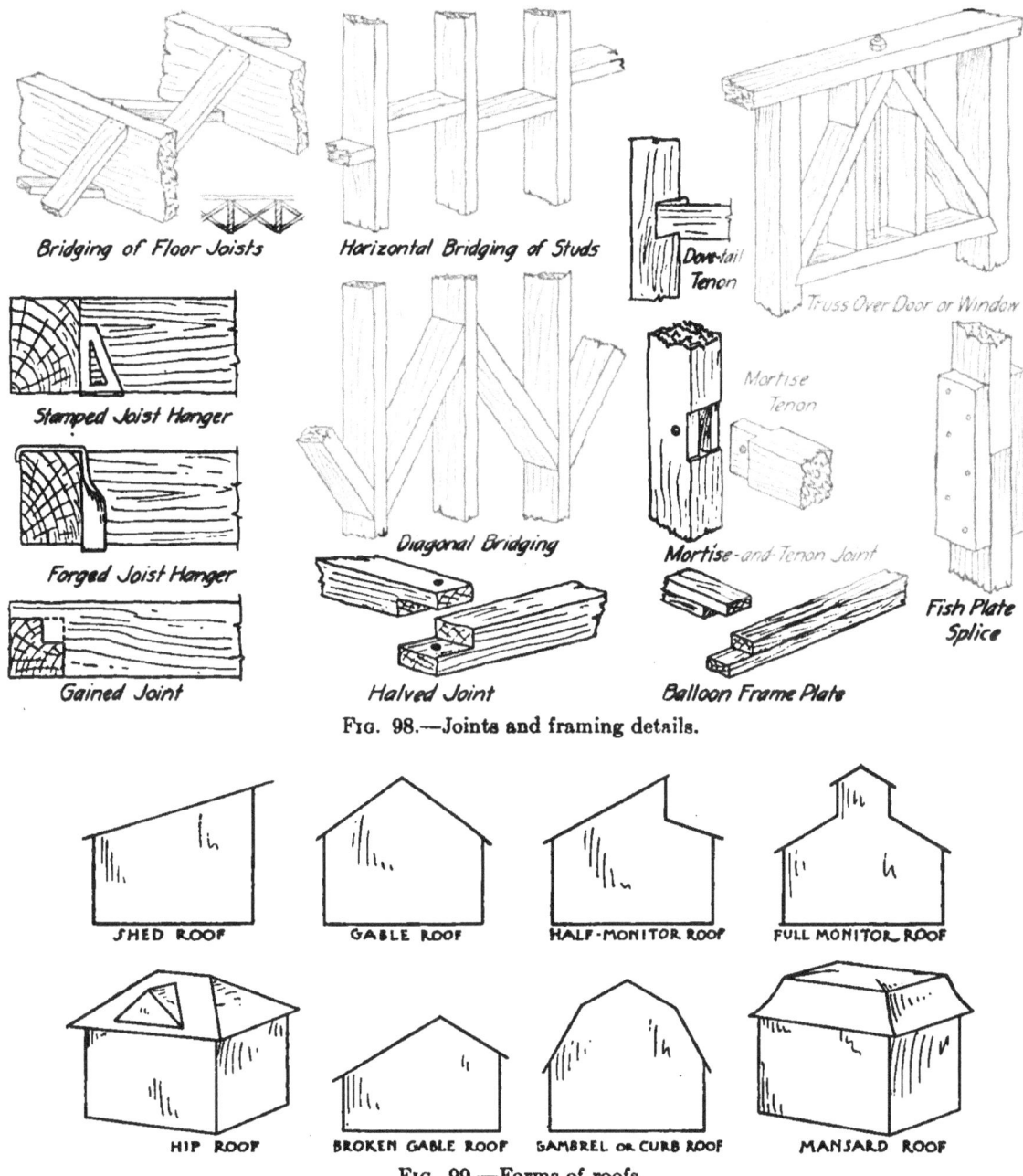

Bridging of Floor Joists

Horizontal Bridging of Studs

Dove-tail Tenon

Truss Over Door or Window

Stamped Joist Hanger

Mortise Tenon

Forged Joist Hanger

Diagonal Bridging

Mortise-and-Tenon Joint

Fish Plate Splice

Gained Joint

Halved Joint

Balloon Frame Plate

Fig. 98.—Joints and framing details.

SHED ROOF   GABLE ROOF   HALF-MONITOR ROOF   FULL MONITOR ROOF

HIP ROOF   BROKEN GABLE ROOF   GAMBREL or CURB ROOF   MANSARD ROOF

Fig. 99.—Forms of roofs.

that it is constructed at some saving in timber (perhaps 10% of complete cost of barn). A greater advantage in cost is in mow space partially free from timbers in pin-joint construction, it is possible to get a mow practically free from obstruction with the plank frame, or the self-supporting roof of the balloon frame.

In deciding upon the kind of framing to be used these advantages and disadvantages must be weighed and local conditions taken into account.

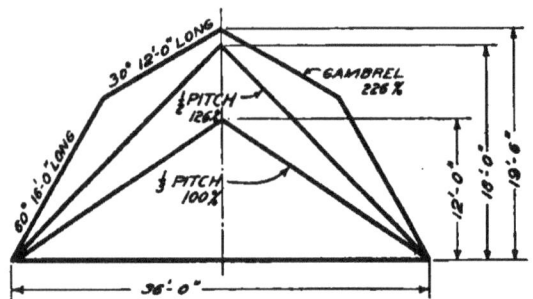

Fig. 100.—Comparative diagram of roof shapes.

In smaller structures the stresses from loads and wind pressure are not so large a factor, and the framing becomes a much simpler problem than that of the barn.

A few examples of the more common

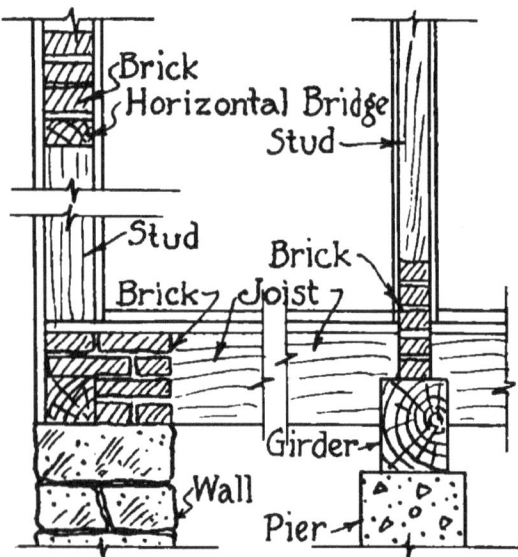

Fig. 101.—Fire-stopping and rat-proofing.

joints and details of construction used in framing are illustrated in Fig. 98.

Fig. 99 shows some of the different forms of roofs used. The pitch, or slope of a gable roof as usually spoken of, is the ratio of the rise to the span, thus when the height of the ridge is one-half the span the slope, evidently 45°, is called "one-half pitch."

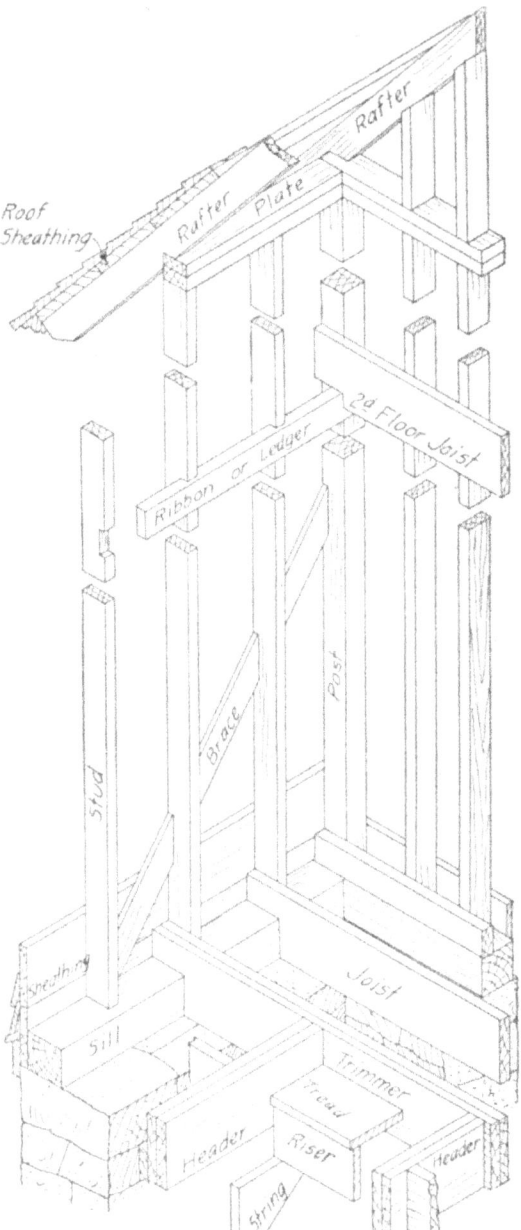

Fig. 102.—Detail of frame construction.

Fig. 100 is a diagram showing the comparative mow capacity of a ⅓ pitch and ½ pitch gable roof, and a gambrel (or curb) roof of the same span.

Fig. 101 illustrates details of fire-stopping and rat proofing, which are important considerations in plank frame construction, particularly so in granaries and residences.

If not already familiar with them, the student should learn the builder's names of the various pieces used in building, as joists,

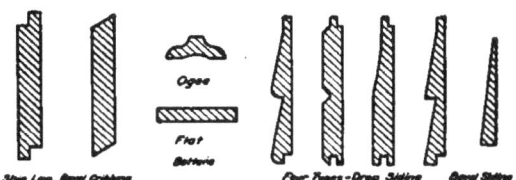

FIG. 103.—Forms of siding.

studs, sills, girders, ribbons, plates and rafters. These names are given on Figs. 102 and 115 and are used throughout the chapter.

The frame of a wooden structure is enclosed by covering it in various ways, by siding or weather-boarding. Different forms of weather boarding are shown in Fig. 103.

structures is occasioning a steadily increasing use of concrete as a building material. It is a material practically indestructible, and with good design and careful supervision makes a most desirable and easily erected structure.

Concrete is made by mixing cement, sand, and gravel or crushed stone in proportions to suit various classes of work. It is put in place by pouring or tamping it into previously prepared forms or molds.

Portland cement as manufactured by the leading companies, is a reliable material, although for important engineering structures every shipment is tested. Poor storage, that is storage in damp places, will injure the best of cements. It is thus well to examine it to be sure that no lumps are present. If lumpy cement must be used, throw out all lumps not easily crushed with the fingers.

Sand should be free from vegetable and earthy matter, although a small percentage of clay is not injurious. In general it should

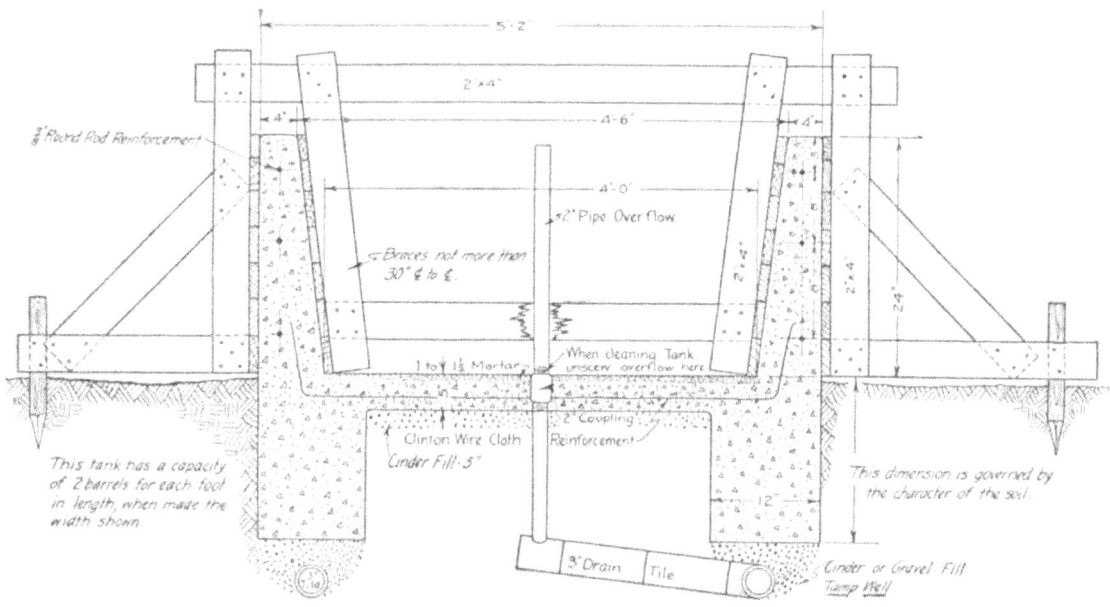

FIG. 104.—Concrete water tank.

Concrete.

The growing scarcity of timber and the increasing demand for fireproof and sanitary

be rather coarse, or a mixture of fine and coarse particles. Gravel and crushed stone should be fairly clean and varied in size.

When concrete is used in beams, tall posts, unsupported floor slabs, or walls sustaining pressures, it has to be reinforced with steel or iron rods, wires or wire netting. The advice of a competent engineer should be sought for such structures. Concrete is an excellent material when properly used, but the large number of failures recorded shows that one can not be too careful in mixing and applying it.

The table given on page 115 will serve as a guide as to the proportions to be used in the various forms of construction, and also the amounts of cement, sand and stone in each cubic yard.

Two distinct things must be considered in drawing concrete structures, first, the representation of the structure itself, and second the construction of the forms. This latter is often a separate problem as the forms must be designed so as to be set up practically and economically, and be strong enough to support the weight of the concrete until it sets. Thus the drawings for concrete structures are somewhat different from those of any other kind.

The section of a water tank shown in Fig. 104 is an example. Notice the use of the symbol for concrete in this and the following figures.

Brick.

One of the oldest known building materials, brick has had and will continue to maintain a preeminent place for durability and beauty. It is used throughout the entire country, but more extensively in sections where brick works are located, as the cost of transportation greatly increases the expense of the material.

In the last few years there has been a great advance made in the quality and texture, or appearance, of face brick and there is more beautiful brick work now being built than ever before. Incidentally, the smooth pressed brick formerly used for fine

work has gone entirely out of vogue, being replaced by rougher texture face bricks of a

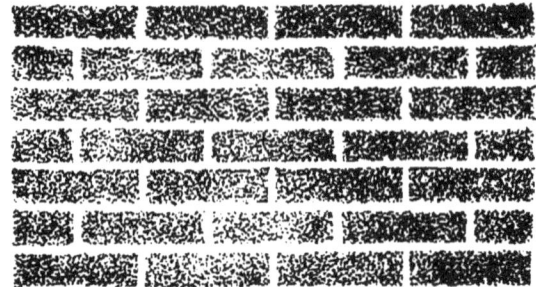

Fig. 105.—American bond.

great variety of surfaces and colors, giving much more artistic and beautiful effects.

The prospective builder expecting to use

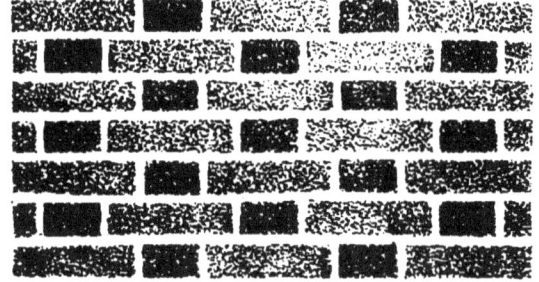

Fig. 106.—Flemish bond.

brick, will inform himself in regard to the kinds available in his section, and the prices of common brick and face brick. In draw-

Fig. 107.—English bond.

ing brick structures there are several points with which he should be familiar. The symbols for brick in section and elevation are shown in Fig. 96. The sizes of brick vary considerably in different localities and for different kinds, an average size is 2¼ × 4 × 8. For careful designing he should

Fig. 108.—Dutch bond.

know accurately the size of brick to be used. Brick walls are drawn 8″ and 12″ thick, or as they are usually termed, 9″ wall and 13″ wall. Face brick are laid usually in

Fig. 109.—Random and coursed rubble.

common or American bond, as illustrated in Fig. 105. Other forms are Flemish bond, Fig. 106, English bond, Fig. 107 and Dutch bond, or as it is sometimes called,

English cross bond, Fig. 108. These form pleasing variations from the American bond but cost a little more for laying. Where resistance to heat is necessary, fire brick should be used.

Stone.

Stone is not so generally common as a wall building material, except in certain comparatively small sections where local stone is plentiful, and even for foundations it is being largely replaced by concrete. Sometimes it is used for picturesque effect, field stones often being used for the purpose.

Stone sills and caps are in general use, and should be dimensioned to come out even with the brick courses.

Stone walls are from 16″ to 24″ in thickness. Fig. 96 shows the symbols used for stone in section and elevation, and Fig. 109 illustrates different forms of laying up.

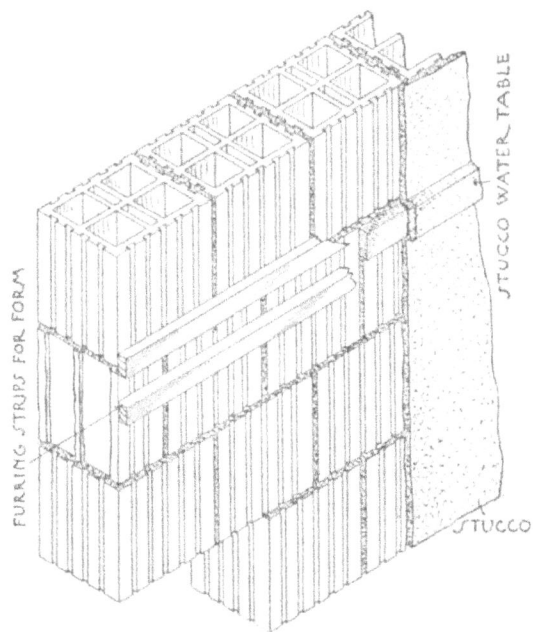

Fig. 110.—Stucco on hollow tile.

Stucco.

Stucco is a mixture of cement and sand with a little lime, applied as plaster to the exterior of a building and finished either smooth or oftener "rough-cast." It is sometimes put on wood lath, oftener on metal lath, and best of all on hollow tile. As a building material it is becoming increasingly popular on account of its appearance and durability. Old brick buildings are sometimes resurfaced by adding a coat of stucco, as it adheres perfectly if properly applied.

Hollow tile with stucco finish is undoubtedly one of the coming building materials. It makes a dry wall, is light, fireproof and easily constructed. A sectional view of hollow tile construction (from the National Fireproofing Company) is shown in Fig. 110.

Roofing Materials.

Wood shingles are more commonly used than any other roof covering. They are made of cypress, redwood and cedar, and various color effects may be obtained by the use of shingle stains.

Slate makes a desirable roofing material. Various colors and grades are on the market. It is heavier than shingles, hence the roof must be framed to carry it. The cost is somewhat greater than for shingle roof.

Roofing tile is used principally for large buildings. It is heavier per square than slate, costs more, and requires sheathing and paper under it.

Composition roofing of various kinds and trade names is very common. It is easily applied and the better grades are very durable.

Galvanized roofing, made of galvanized iron or steel is coming into general use. It comes in lengths 6, 8, 10 and 12 feet, and the framing should be built to suit. It has the advantage of being cheap and easily put on, but would not be used on residences on account of its appearance and its heat absorption.

PREPARING PLANS

In preparing plans for any building the designer will go through the following operations:

(1) Make preliminary sketches.
(2) Carry through the general drawings, plans, elevations and sections.
(3) Make detail drawings of such parts as require additional explanation.
(4) Take off the quantities required of each material and tabulate in a bill of material.
(5) Write specifications.
(6) Estimate the cost.

To illustrate these steps, the design of a dairy barn, including drawings, specifications and bill of material has been carried through, commencing on page 57, and will serve as a guide for other buildings.

DAIRY BARN

With the sale of milk regulated by health authorities, and the successful attempt to stop filth at the source, the present-day dairy barn has come to be a clean, well lighted and well ventilated structure.

The location must be well drained, or so situated as to make good drainage possible. The barn should be faced if possible so that the yard is at the south or east. Ample storage for feed should be provided, and bedding must not be overlooked. Provision must be made for the removal of manure at least once daily, to a pit required to be at least 50 feet from the barn, but better if placed 100 feet away. The floors should be of some water tight, non-absorbent material. Concrete fills this requirement, but is objected to by some persons. The objections may be overcome by making the floors with sand or float finish, and by flooring the stalls with cork brick or creosoted wood blocks, or by using a plank overlay in them over the concrete.

The lighting should be well distributed, and at least 4 feet of glass provided for each animal. There should be at least 500 cubic feet of space allowed for each animal and in addition flue ventilation should be installed of such capacity that each animal will have 25 square inches of flue area.

to prevent dust, dirt and seed from sifting through. Hog houses or chicken houses must not be located within 100 feet of the barn.

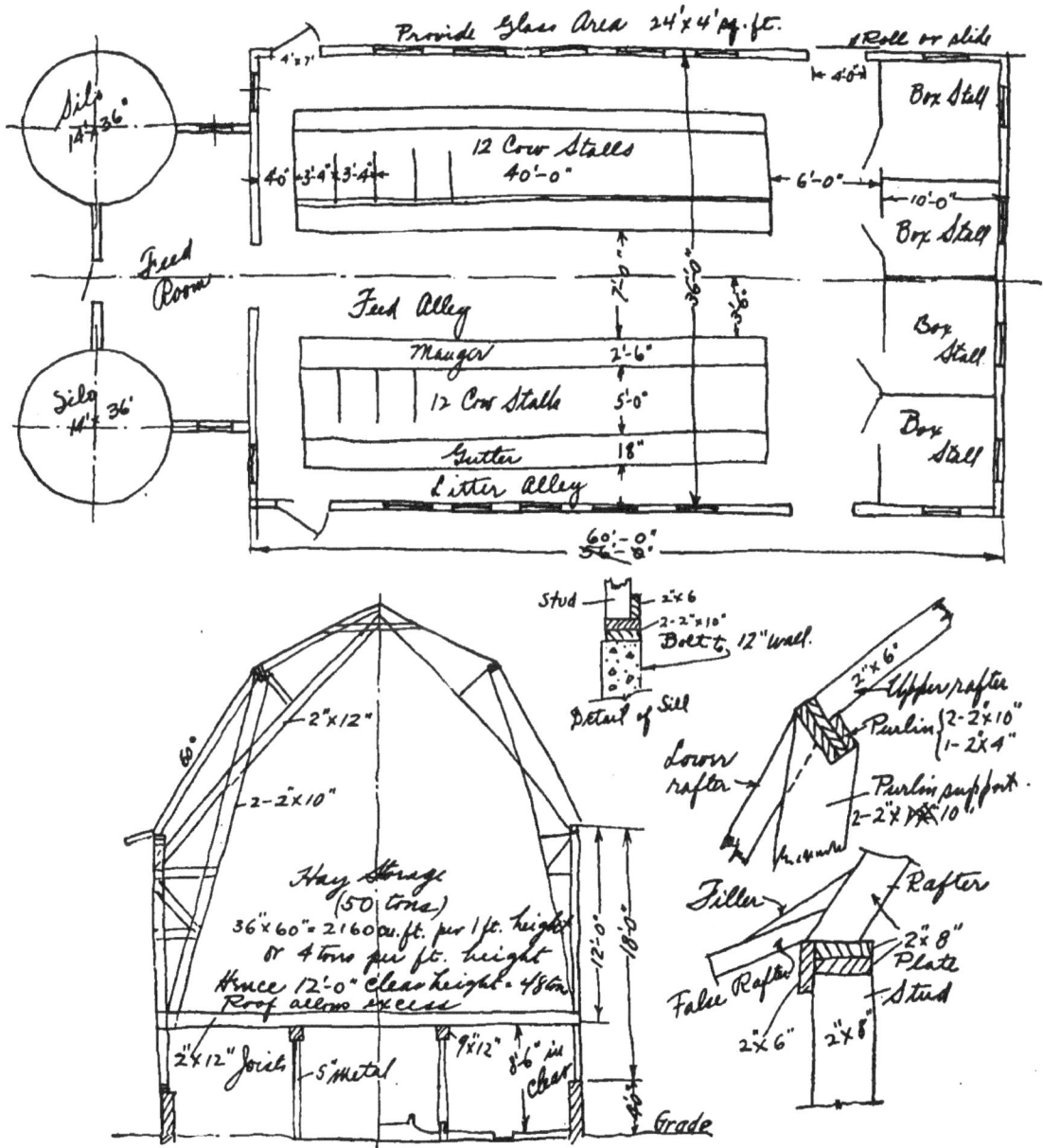

FIG. 111.—Preliminary sketch of dairy barn.

Flues with a cross sectional area of less than 200 square inches are not very efficient.

Where hay or straw is to be stored above the stable, the floor must be tightly ceiled

In planning, the following data may be used. Small cows require stall space 3 ft. wide, 4'-6'' long; large cows from 3'-6'' to 4'-0'' stall width and 5'-0'' to 5'-6'' in length.

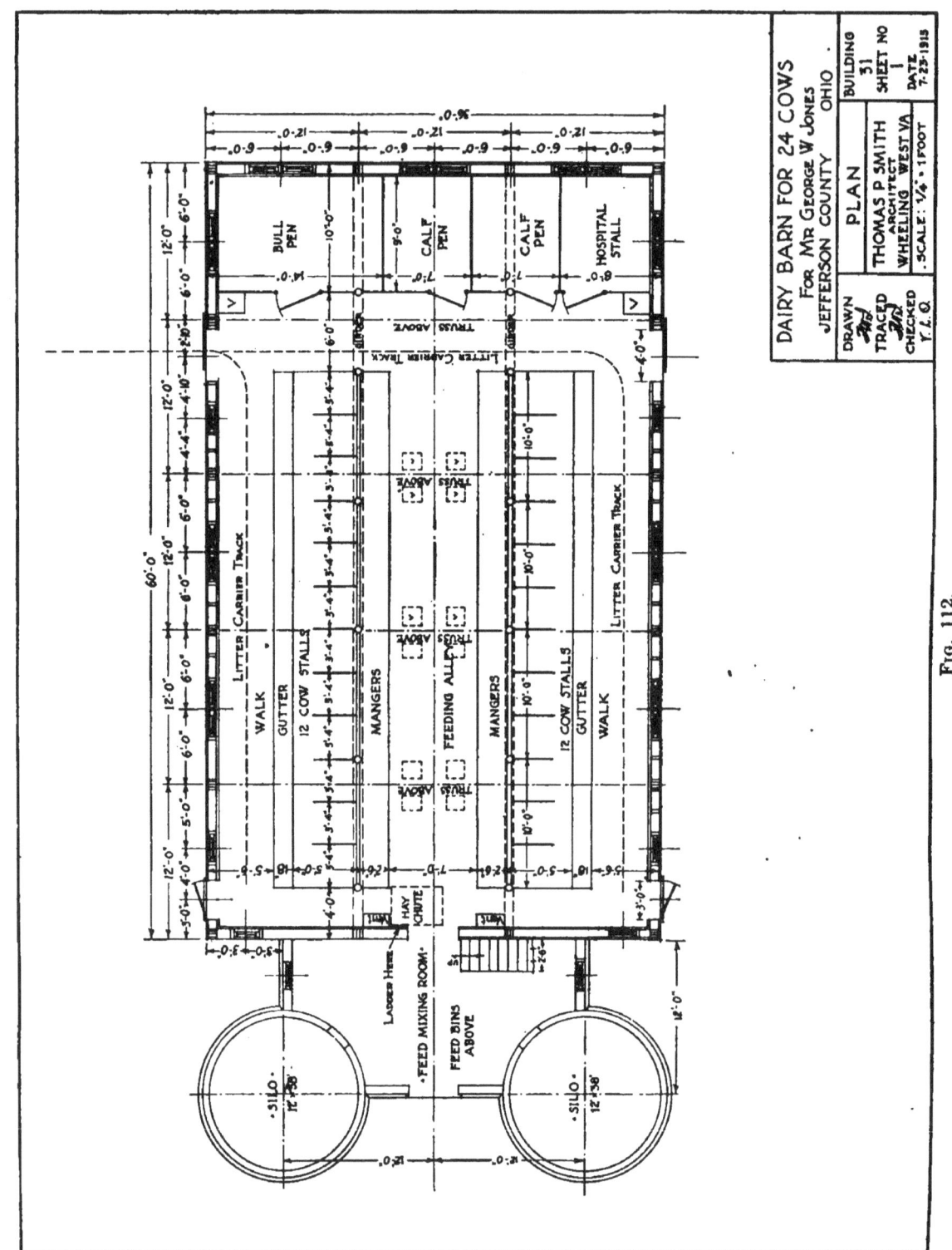

Fig. 112.

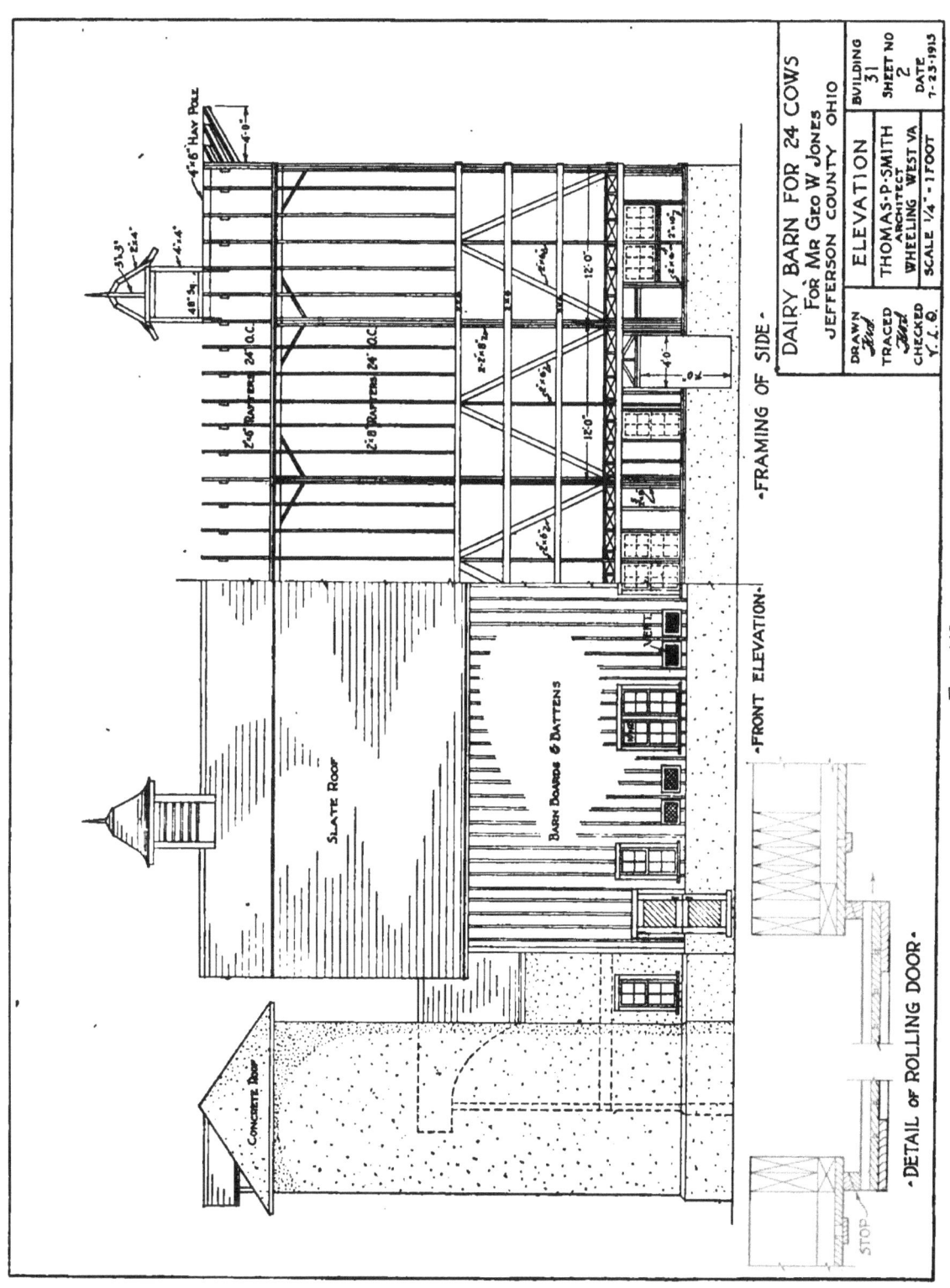

FIG. 113.

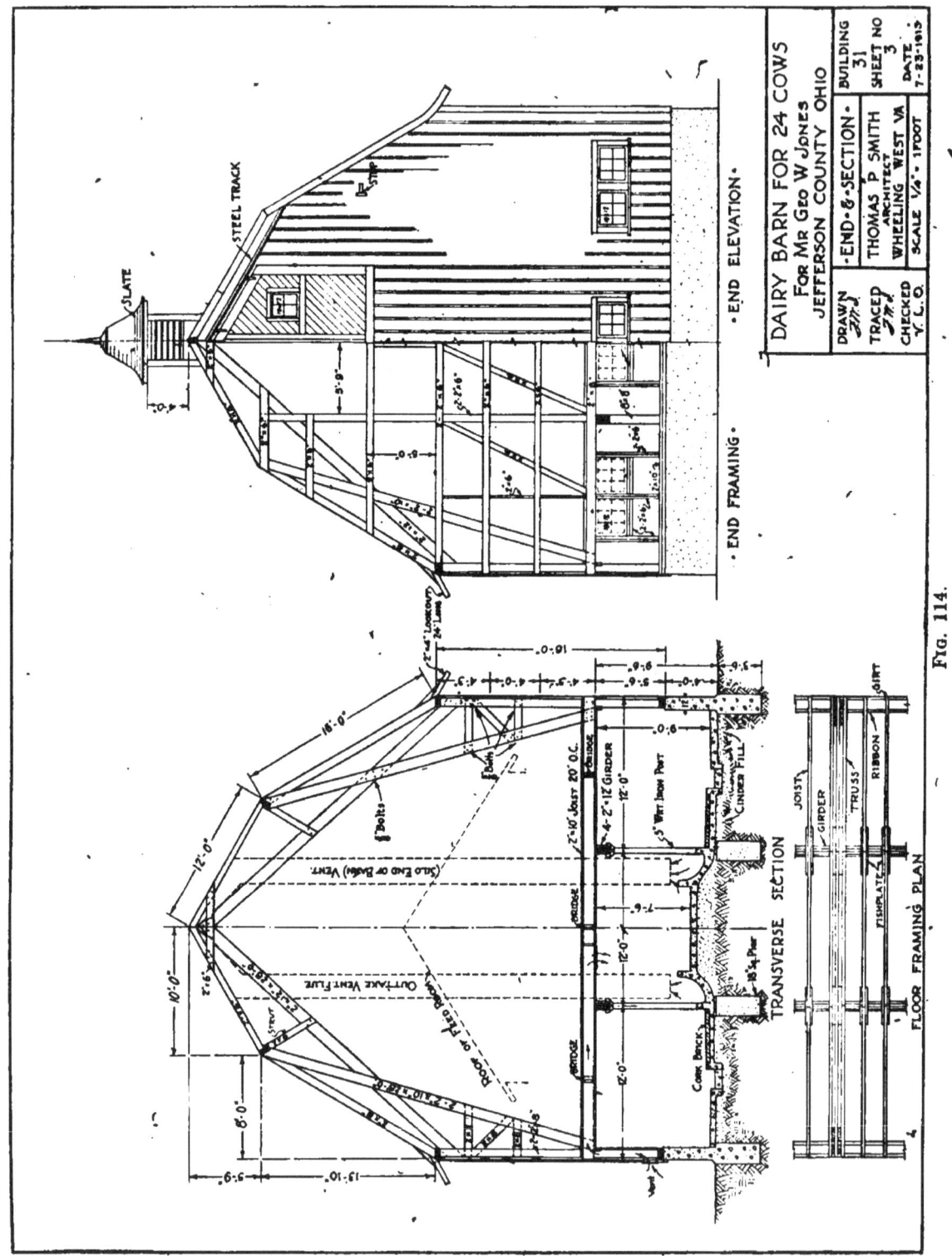

FIG. 114.

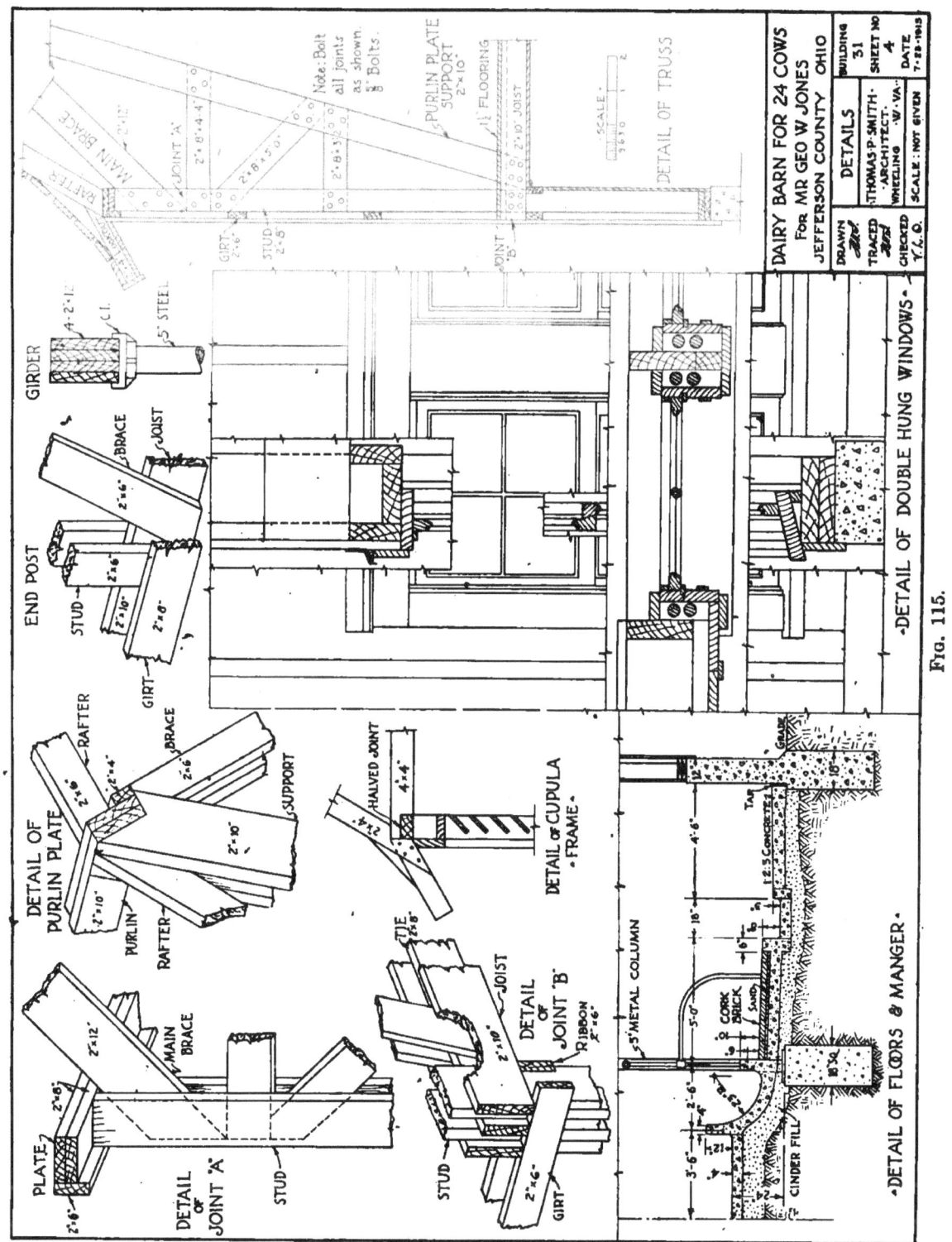

Fig. 115.

Mangers may be from 2'–6" to 3'–0" wide. Feeding alleys should not be less than 3 feet wide. Litter alleys should have a minimum width of 4 feet, and this increased to 8 feet if a spreader is to be driven through. Gutters may range from 12" to 22" in width and vary from 5" to 16" in depth. An average gutter is shown in detail in Fig. 115.

Floors and stalls should slope toward the gutter from ⅛" to ¼" per foot. A slope of 1" in 10 feet will be sufficient for the flow of the gutter.

In figuring storage space, calculate on the average basis that each animal will require the following amounts per feeding season: 4 tons of ensilage, 2 tons of hay, 1000 lbs. of grain or other concentrates, and 2 tons of straw bedding. Data regarding the space occupied by various materials is given in Chapter VII.

A careful study of the dairy score card given on page 116 is advised. This may be used in checking up a set of plans, or in scoring an existing barn to see if it conforms to the requirements. Different states have similar score cards, which may be obtained from the State Dairy and Food Commissioner.

Plans for a Dairy Barn.

Beginning with Fig. 111 illustrating the preliminary freehand sketch, a complete set of drawings for a dairy barn of modern design to accommodate 24 Holstein cows and providing for bull, calves, and hospital cases, is shown. Silos and hay capacity have been estimated for a year's feed.

Fig. 112 is the plan of this barn, Fig. 113 the side elevation, and Fig. 114 the end elevation and sections, showing that the construction is the plank truss type. The plank truss frame is usually called the "Shawver barn," as it was first developed by Mr. John L. Shawver.

Fig. 115 is a sheet of details showing floor and manger construction, window detail and framing joints. This sheet, incidentally, illustrates the advantageous use of pictorial drawing for details of construction.

Specifications.

Accompanying the drawings for any structure should be the specifications, containing a definite, · itemized statement regarding the materials and methods of construction required. In a very simple structure it is possible to include all of this information as notes on the drawing, but in such a structure as a barn or residence, they should be typewritten, under a separate cover and the different items grouped under appropriate headings.

After the introductory head, specifications always contain first, what is known as the general condition clause, covering all the work and forming part of the contract. This clause should therefore be worded carefully so that in case of dispute, its meaning can not be questioned.

The general condition clause is followed by specific clauses for the different classes of work and material, usually about in the order of their occurrence in the progress of construction.

The following specifications for the dairy barn illustrated may serve as an example of the general form used for such structures. These are short and concise, but sufficient for full explanation of the requirements. Architects' forms for residences and public buildings are much longer, particularly in the general condition clause.

Specifications for Dairy Barn.
Specifications of materials and workmanship required for the erection and completion of a frame dairy barn for Mr. George W. Jones, Jefferson County, Ohio.   THOS. P. SMITH, .
                                        Architect.

General.
All materials shall be of best grade and no substitutions shall be allowed in place of material hereinafter specified, except on written order of the

owner and architect. All work must be done in the best possible manner by skilled workmen. Work not done according to the manner specifically stated, must at the discretion of the architect, be taken out and properly constructed, all expense attached for such work and materials to be borne by the contractor.

All rejected material must be removed from the premises within 24 hours after notification.

Any changes from the drawings or specifications, or any extras, are to be agreed upon in writing before being made.

Excavation.

The excavation shall be made according to lines and levels set by the architect, and all trenches dressed smooth. Tile drains shall be laid as directed by the architect, tile being furnished by the owner. The laying of such tile is to be paid for by the owner at a rate not to exceed 30 cents for each lineal rod.

Concrete.

All concrete shall be composed of Portland cement, clean sand, and crushed stone ranging in size from $\frac{1}{2}''$ to $1\frac{1}{2}''$. The proportions shall be as follows; for all piers and walls below grade, 1 part Portland cement, 2 parts sand and 4 parts stone. All floors, mangers, gutters and other concrete work not included in piers, foundation and walls, shall be one course work with 1 part cement, 2 parts sand and 3 parts stone. All cement, sand and stone will be inspected and approved by the architect. A batch mixer must be used and each batch turned not less than 25 times after the addition of water. The mixture must be wet enough to fill all forms and trenches and always show a film of water after reasonable tamping.

All forms shall be water-tight and made of material strong enough to prevent bulging. Forms for work above grade must be of dressed lumber and leave the walls smooth.

All concrete shall be placed within 20 minutes after mixing, and tamped free from voids.

Retempering of concrete will not be permitted.

Timber.

All timber shall be well seasoned. All framing lumber shall be clear grade, long leaf yellow pine free from shakes and winds. Siding shall be clear cypress (red cedar, redwood or white pine). Mow flooring, tongue and grooved yellow pine (hemlock, fir or local wood) laid tightly in painted joints. Roof sheathing to be No. 1 hemlock laid close and covered with good quality tarred roofing felt,

weighing 40 lbs. per square. All trim and finishing lumber to be clear white pine.

Slate shall be No. 1 grade of (black Bangor) slate.

Gutters, spouting and flashing shall be No. 24 gauge galvanized iron, painted with two coats red lead and linseed oil paint. All exterior wood work shall be primed with pure boiled linseed oil and best quality white lead. Two additional coats of pure linseed oil and pure white lead paint shall be applied as soon as consistent with good results.

All hardware and stable fittings to be furnished by the owner, and set in place by the contractor.

BILL OF MATERIAL

A complete bill of all the materials required in the construction would be needed for ordering and estimating. Concrete piers and walls are figured for their cubic contents, and reduced to cubic yards. Concrete floors are estimated in square feet (or square yards), excavation in cubic yards, lumber in board measure, slate, roofing felt and paint, in squares (100 sq. ft.) and hardware, stable fittings, etc., from dealers' prices obtained. Lumber should be figured in length to the nearest even foot over, and in tongue and grooved flooring and ceiling $\frac{1}{8}$ more than the total should be added.

The bill of lumber of the superstructure of the dairy barn is here given as a guide.

Bill of Materials, *Superstructure.*

Sill....2 × 10 − 384 linear ft. = 640 ft. B.M. y.p.
Girders 2 × 12 − 480 linear ft. = 960 ft. B.M. y.p.
Posts ..4 − 8 × 8 × 4′-0″ = 86 ft. B.M. y.p.
                                                  or oak.

6 *Trusses*

	Size	No.	Length	Ft. B.M.	Material
Posts.........	2 × 8	24	18′-0″	576	y.p.
Main Brace...	2 × 12	12	28′-0″	672	y.p.
Purlin Support	2 × 10	24	28′-0″	1120	y.p.
Short Brace...	2 × 8	12	14′-0″	224	y.p.
			(cut into 4′-4″, 5′-0″ and 3′-2″)		
Foot pieces (Foundation & Ribbon).....	2 × 8	36	6′-0″	288	y.p.
Purlin Strut...	2 × 6	24	4′-0″	72	y.p.
Purlin Braces..	2 × 6	22	8′-0″	176	y.p.
Collar Beams..	2 × 6	31	6′-0″	186	y.p.

	Size	No.	Length Ft.	B.M.	Material
Intermediate Posts........	2 × 6	18	18'-0"	324	y.p.
Inter. Posts (above end plate)......	2 × 6	8	14'-0"	112	y.p.
Inter. Braces..	2 × 6	28	14'-0"	392	y.p.
Girts (Nail). .	2 × 6	68	12'-0"	816	y.p.
	2 × 6	4	14'-0"	56	y.p.
	2 × 8	6	12'-0"	96	y.p.
Ribbon......	2 × 6	10	12'-0"	120	y.p.
Joists........	2 × 10	118	12'-0"	2360	y.p.
Joists at truss.	2 × 8	18	12'-0"	288	y.p.
Bridging......	1 × 3	....	432' lineal	108	y.p.
Rafters (short)	2 × 6	62	12'-0"	744	y.p.
" (long)	2 × 8	62	16'-0"	1224	y.p.
Lookouts & filler........	2 × 4	....	166'-0" lineal	112	y.p.
Hay Pole......	4 × 6	1	16'-0"	32	y.p.
Plate.........	2 × 8	20	12'-0"	320	y.p.
Purlin......	2 × 10	20	12'-0"	400	.y.p.
	2 × 4	10	12'-0"	80	y.p.
Studs (short at lower floor)...	2 × 6	42	12'-0"	504	y.p.

Covering { 1 × 12 × 18'-0" 3800 ft. B.M. } cypress
{ 1 × 12 × 16'-0" 800 ft. B.M. }

Battens... 1 × 3 3896 lineal ft. 975 ft. B.M. cypress

Roof Sheathing—3840 ft. B.M. 1" × 10" random length.

Slate—16" × 10" 39 squares.

Windows—

14 frames complete for 10" × 12". 8 Light check rail windows, 4 double as per details on Sheet 4, 6 single windows. 56 sash wts. C.I.

14 check rail windows 8 Lt. 10" × 12".

4—4 Lt. barn sash 10" × 12" for gable.

10—6 Lt. barn sash 10 × 12 complete with frames

24 lineal ft. track and hangers for gable doors.

400 lineal feet 1 × 6 white pine trim.

2 cupolas to be built at mill.

Floor—2400 sq. ft. superficial area, 1½" × 6" T. & G.

Ceiling 2400 sq. ft. 1 × 4 T. and G.

Estimate.

The estimated cost of this barn, without silos and feed room, is $4200. One thousand dollars, as an average estimate, may be added for these items.

Data used as a basis in determining the estimate will be found under "estimating," on page 119.

THE HORSE BARN

The essentials for a good horse barn are, convenience of location, good drainage in all directions, convenient interior arrangement, plenty of light, and ample storage for hay and grain.

The walls may be of a single thickness of barn boards, if they are tight and draft proof. The floor may be of plank, tamped clay, concrete or cork brick. Clay is sometimes considered as the best floor, but concrete has proven to be satisfactory and sanitary; usually however, a removable plank overlay is used in stalls in connection with the concrete floor. Brick with plank overlay is also used sometimes. Cork brick makes an ideal floor, being water proof and resilient, but the initial cost is high (about $55.00 per thousand). It should be laid on concrete foundation.

The ceiling should be at least eight feet high in the clear, and nine feet would be a better height. Seven hundred to one thousand cubic feet of space should be allowed for each animal.

Single stalls are made from four feet to six feet in width. Five feet is comfortable for a heavy horse. Double stalls are usually eight feet wide. The length of stalls is nine feet, six inches including mangers. Box stalls range in size from 8 × 10 to 10 × 12 feet.

An average manger is two feet wide and three feet six inches high. The side toward the animal should slope inward from four inches to six inches, to prevent injury to the knees.

Feed alleys, if used, should be at least three feet wide. The litter alley should be at least four feet wide, to allow a medium horse to back from the stall comfortably. If a manure spreader is to be driven through, eight feet will be necessary.

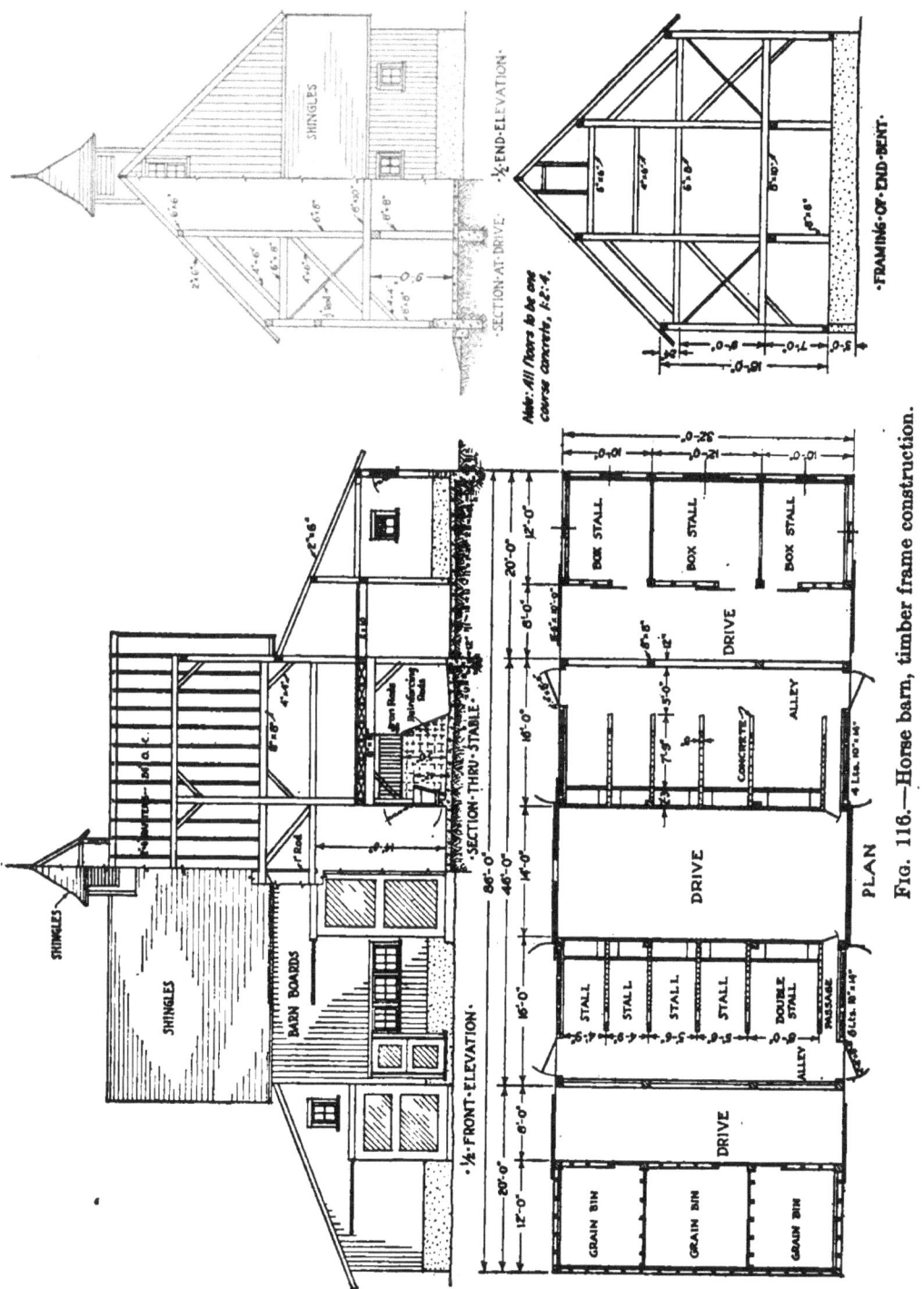

FIG. 116.—Horse barn, timber frame construction.

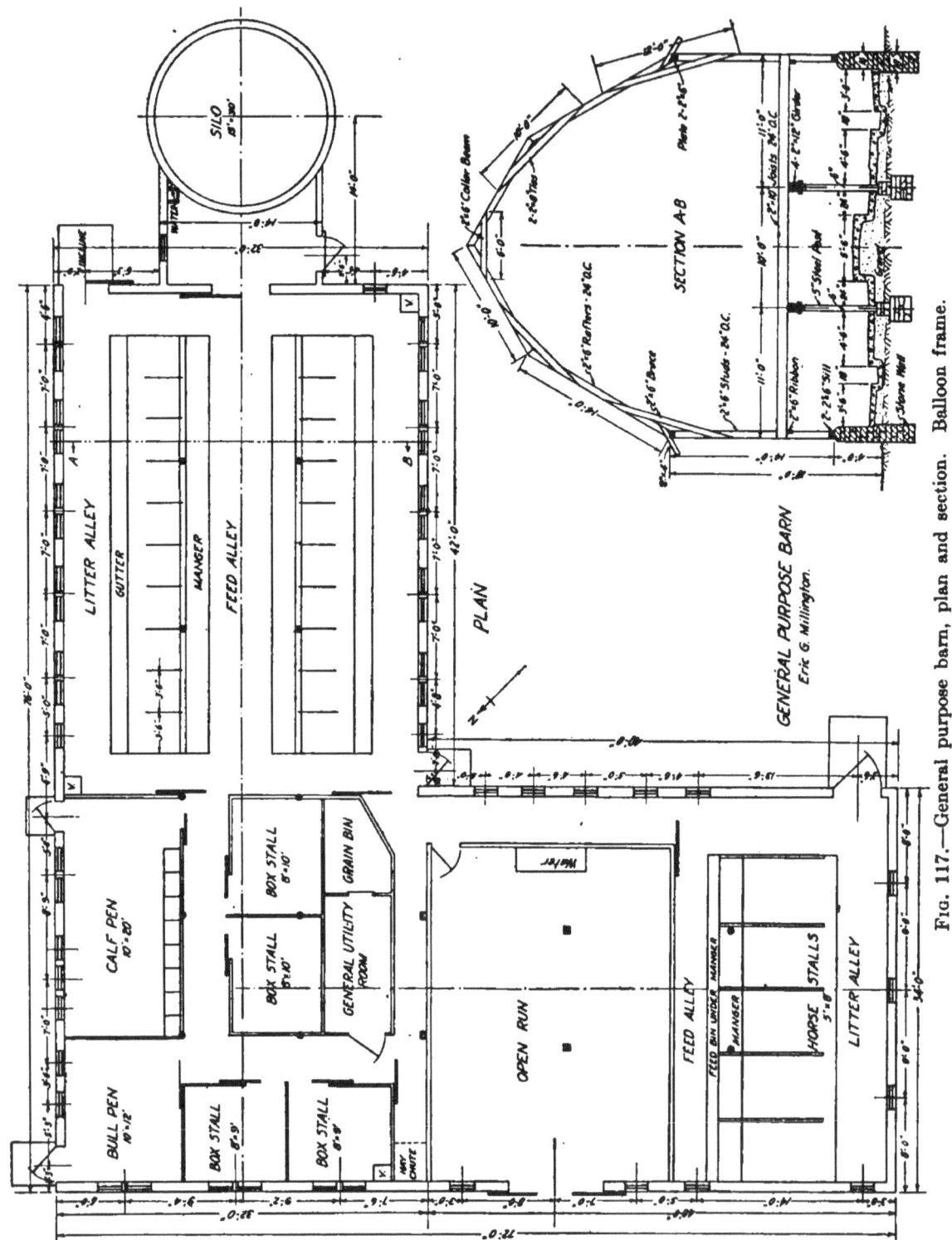

Fig. 117.—General purpose barn, plan and section. Balloon frame.

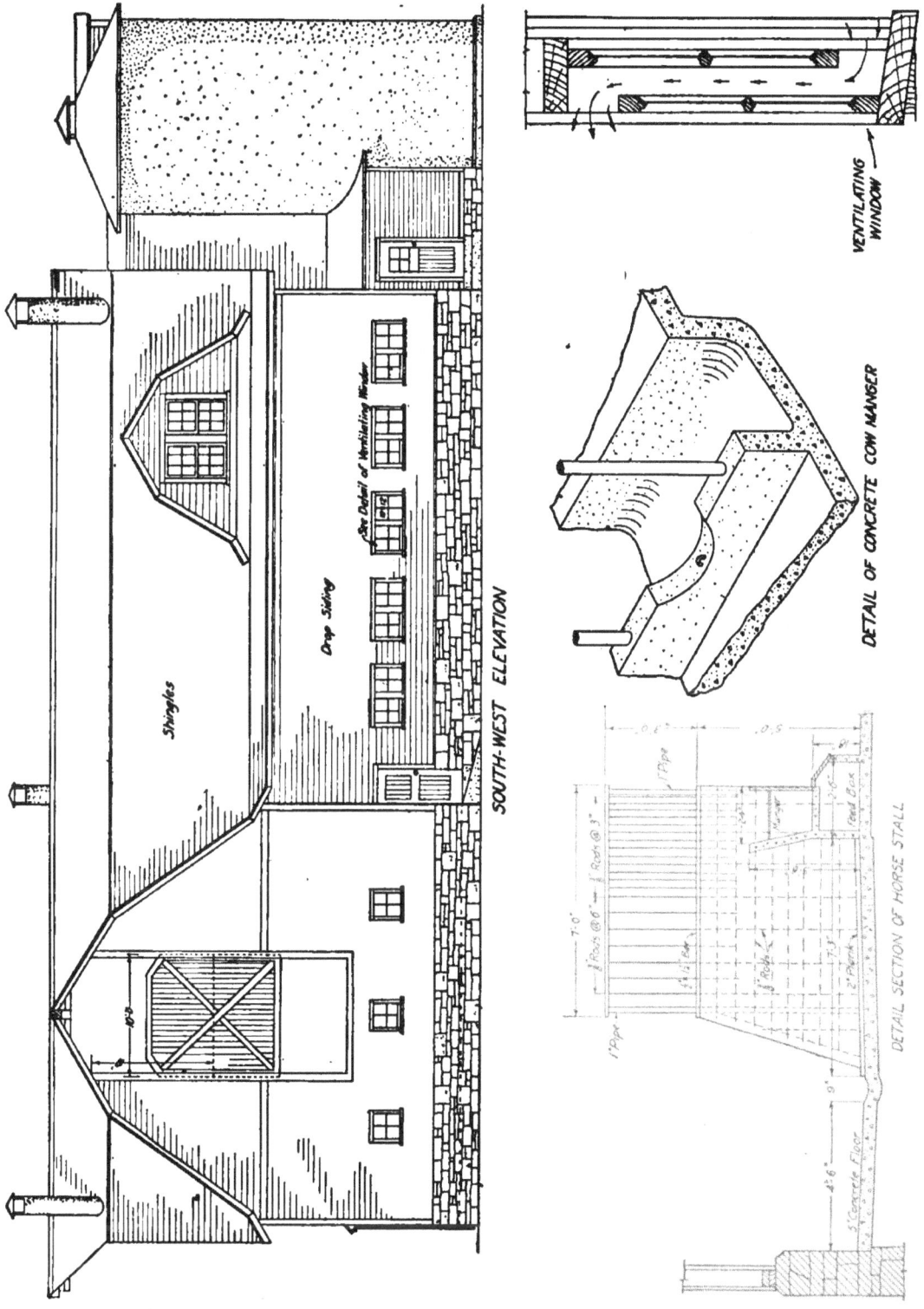

FIG. 118.—General purpose barn, elevation and details.

Windows with sash tilting inward at the top, make a good means of ventilation. When ventilating flues are used, thirty square inches should be provided for each horse.

Stall partitions should be very solidly constructed for at least five feet in height. Some prefer to build them six feet. A five foot solid partition surmounted by steel bars or heavy wire mesh has the advantage that it allows the horses to see each other and does not shut the light from interior stalls.

The horse barn shown in Fig. 116 may be studied as an example of the application of the features mentioned in a compact and practical design. The plan is a familiar one in that the horses are fed from the main drive-way. Solid drop swing doors are provided on the manger fronts so that in cold weather the stable may be closed tightly. The size and spacing of the windows is designed for ample light and ventilation.

A concrete floor extends throughout the length of the barn. In the stable proper the stall floors and alley slope toward a shallow gutter at the rear of the stall. The stalls are floored with plank over the concrete.

The stall partitions are of concrete reinforced with $\frac{3}{8}''$ round rods extending above the partition and forming a grating up to the ceiling. The stalls are of different widths to accommodate horses of varying sizes. Three box stalls are provided.

A detail of a horse stall is shown in Fig. 118.

The frame is the pin joint timber frame type and is constructed so that the mow space is free from cross beams, thus allowing the use of the newer type of hay slings. The space above the drive-way in the lean-to may be used to advantage for the storage of hay or grain. This barn would be built by raising the two longitudinal bents first, this construction permitting a more economical arrangement of floor joists.

THE GENERAL PURPOSE BARN

On farms where the separate highly specialized dairy or horse barns are not required, a combination general purpose barn for cattle, horses, and possibly other farm animals, as well as for the storage of feed, etc., is used. When this barn contains a dairy stable it is necessary to observe the laws in force for dairy barns, with the addition that there must be provided either a tight partition between horse and cow stables, or a clear space of twelve feet between the cows and the horses.

A common form of general purpose barn now in use has attached to it a large storage shed for straw, the first story of which is used as a covered and sheltered barn yard. This barn yard should be paved with concrete or brick. It is estimated that the paving of such a yard will pay for itself within two years in the value of the manure saved.

An example of another form of general purpose barn is shown in Figs. 117 and 118, which give the plan, elevation and section of an L shaped barn of the balloon framed and self-supporting roof type of construction. This barn is designed to face with the internal angle of the L to the south, so as to secure a maximum of sunlight throughout the day. Both the rolling doors and the open run between the horse stable and the milking stalls comply with the law in effectually separating the two stables.

The cow stalls are designed to be used for the feeding of concentrates, and for milking only. Roughage is to be fed in the open run, where the cows are confined except during milking time.

Ample storage space has been provided for hay, straw, grain and silage.

THE SWINE HOUSE

There are two distinct types of swine houses in general use, the individual cot and the colony house.

The individual house is used where one sow and her litter are to be kept apart from other swine. It is usually built upon skids so that it may be moved from time to time for sanitary reasons or to change the run. A common and inexpensive cot is shown in

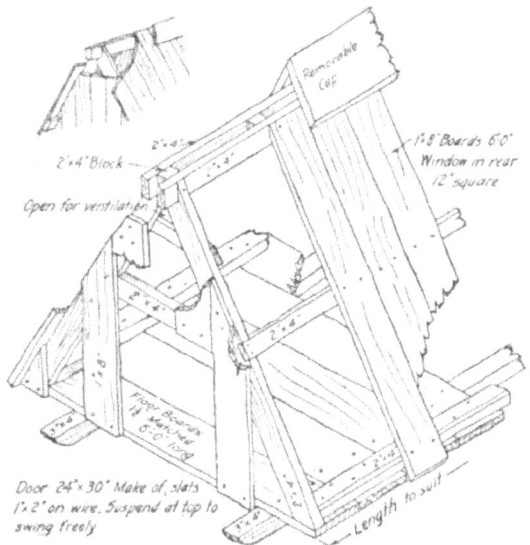

Fig. 119.—"A-type" individual hog cot.

Fig. 119. Fig. 120 shows other models of the same general class.

Fig. 121 shows a unit plan and sectional elevation, together with some details, of a colony hog house. The half-monitor roof shown is peculiarly adapted to give a two row hog house sunshine on both sides of the alley. It will be seen that, to be effective, the house must face with windows to the south. The plan shows two units of 2 pens each. The house may be made of any desired length by adding the required number of units. A good scheme to follow for a large plant would consist of a central feed storage and mixing house with principal axis running north and south, with wings of five units each extending to the east and west respectively. This would make an establishment of 20 pens and feed house that would be very well arranged to save time

and labor in feeding and at the same time practically divide the herd, which might be desirable in time of epidemic.

Sunshine tables are published in Farmer's Bulletin No. 438 U. S. Department of Agriculture, enabling one to build the house so as to take advantage of the sun in his locality during the months from January to June inclusive. An abbreviation of this table will be found on page 117, Chapter seven.

Fig. 122 shows two other types of house, A, with a full monitor roof, which is best when necessary to have the house running north and south; and B, for a house facing south, with a single row of pens.

A *dipping vat* is connection with the hog house may be made 30″ wide at the top and from 18″ to 20″ wide at the bottom. The tank should start with one vertical end, and

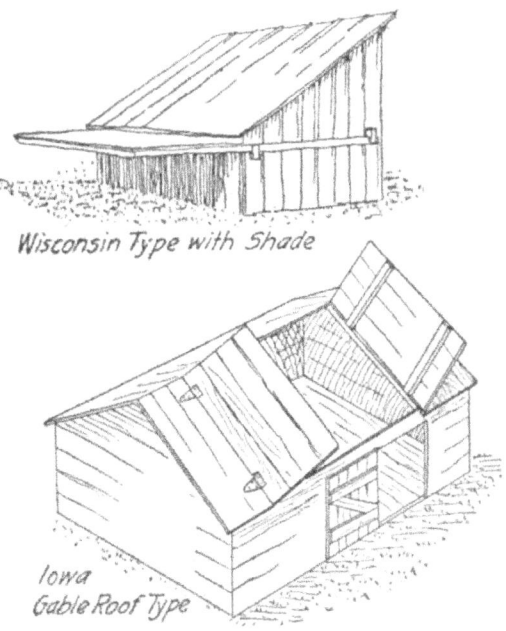

Fig. 120.—Individual hog cots.

continue with a level bottom 3′–0″ deep and 6′–0″ long. A slope 7′–0″ long to the level of the floor should be provided, to aid the hogs in getting out. A tilting board at the

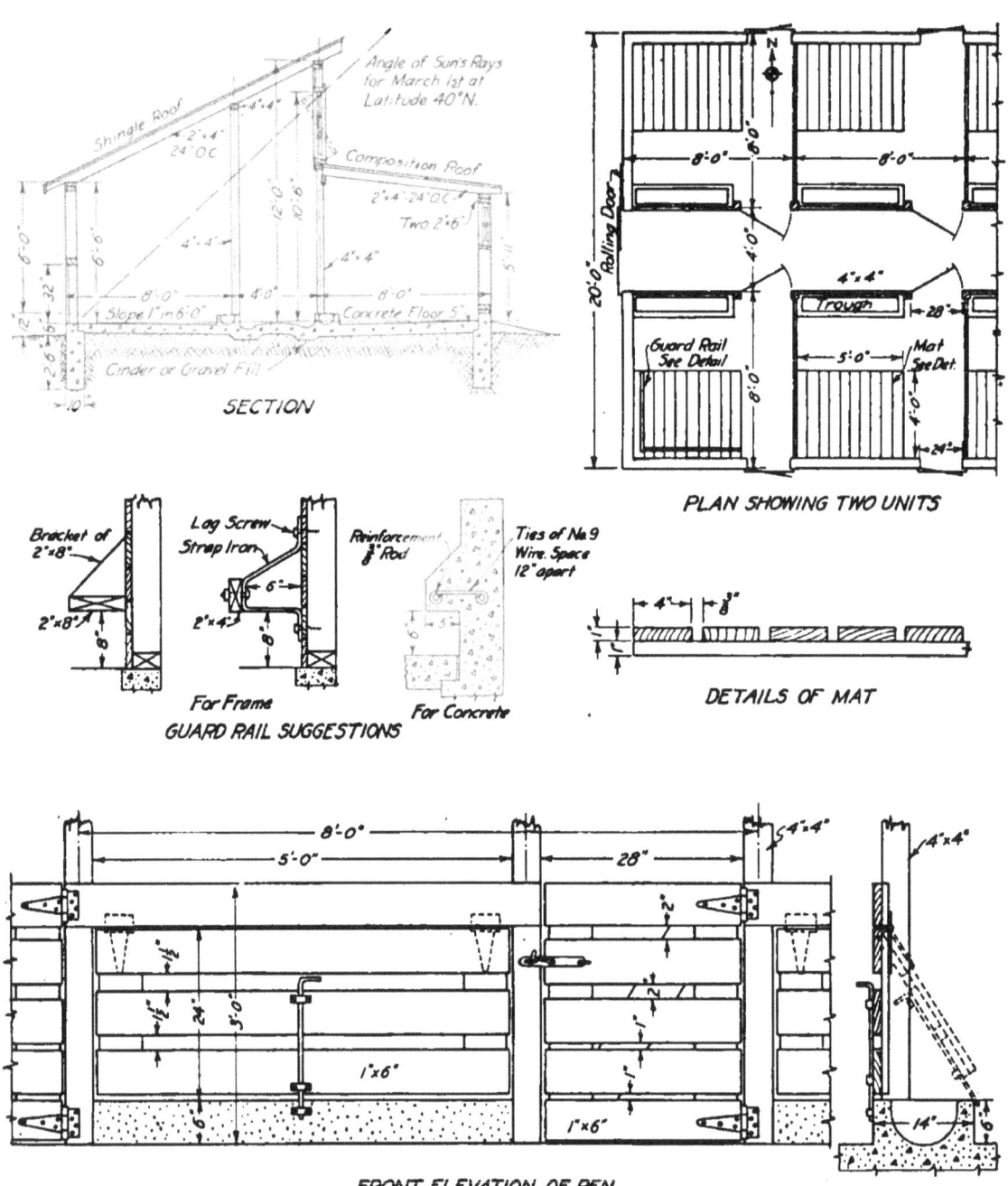

FIG. 121.—Swine house, colony type with half-monitor roof.

perpendicular end is used for throwing the animals into the dip.

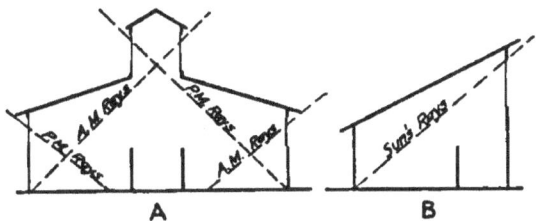

FIG. 122.—Alternate forms of colony swine house.

THE SHEEP BARN

The requirements for sheep sheltering are few so long as the sheep are kept dry and out of drafts, hence the location should be well drained, and while the shed may be open toward the south or southeast, it provided for dividing the space into temporary pens.

Feed storage should be provided for as follows:

2 to 4 lbs. ensilage, 4 lbs. of hay, and 2 to 3 lbs. of grain per sheep per day.

THE POULTRY HOUSE

It is only possible in this discussion to present the general requirements of such a building as a poultry house. There is no generally accepted standard form as compared to other farm buildings, as climatic and other conditions vary so widely in different parts of the country. There are prevailing types in different sections, each well adapted for its particular location.

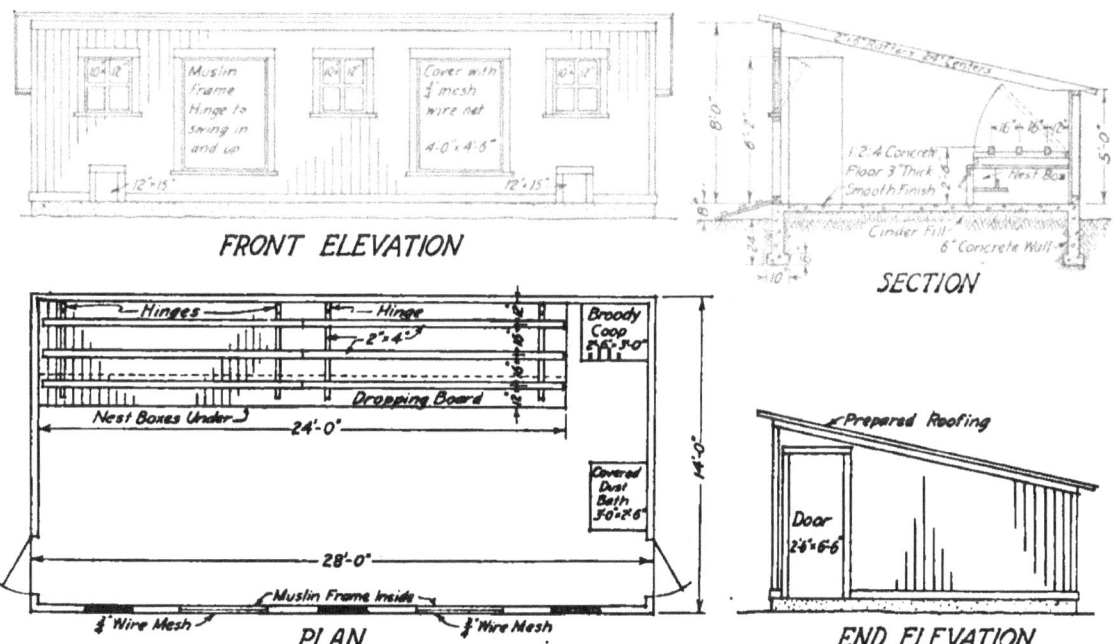

FIG. 123.—Poultry house designed for 100 hens.

should be so built as to permit closing in case of driving storms from the south.

In designing, twelve square feet of floor space should be allowed for each breeding ewe, and six square feet for each fatting lamb.

A sufficient number of hurdles should be

Comfort may be called the key word in poultry house construction. Fresh air is demanded, hence the *ventilation* is an important consideration. Care in both drainage and ventilation will give the *dryness* required. *Sunlight* must be admitted, thus the house should face south or southeast, if possible being built on a south or southeast slope. One square foot of glass area for each twenty square feet of floor space is a minimum, but other openings for ventilation should be provided. These openings should be so placed as to allow circulation without drafts. If covered with muslin frames, the cloth will allow good circulation without any draft, and will give sufficient protection during the cold months. Too much glass radiates an excessive amount of heat, and sometimes causes condensation of too much moisture.

Small houses with a capacity of from seventy-five to three hundred birds are to be preferred, and from two to five square feet of floor space to each bird should be allowed.

Floors may be of earth if well drained, but the concrete floor has the advantage of being sanitary and rat and vermin proof.

Roofs should be water-tight. They may be of any type, as shed, half monitor or gable, but the shed roof with slope to the north is cheap and satisfactory, if the span is not over fourteen feet wide.

Roosts should be about ten inches above the dropping boards, from twelve to fourteen inches apart, and eight to twelve inches of space allowed for each fowl. The roost frame should be hinged to facilitate cleaning the dropping board. Nests must be convenient for the laying hens. One nest for three to five birds is a fair allowance. They should be twelve by twelve by sixteeen inches and placed about twelve inches above the floor. Wire mesh bottoms will aid in cleaning. The top board of the nest box should slope, to prevent roosting, unless the nests are placed under the dropping board. A good trap nest is shown in Fig. 2, Chapter I.

Fig. 123 is a house designed for one hundred birds, and will give suggestions as to possible arrangement and construction.

IMPLEMENT SHEDS

It is generally recognized that he is a wise farmer who houses his implements and tools when not in actual service in the field. Experience has shown that taking into account the amount and kind of machinery now in use, a shed of eighteen to twenty-four feet in width is the most economical in construction. Twelve foot openings will take in the widest machine, and if made twelve feet in height will accommodate tractors, hay-loaders and other implements requiring head room.

The best way to plan a machine shed is to make a list of the implements to be housed, including probable later additions. Measure the overall dimensions, length and width, of each of these machines, lay off these dimensions separately to scale and cut out paper rectangles representing the different machines, writing the name on each. By shifting and arranging these little dummies, bearing in mind the season and use of each implement, the approximate dimensions of the building necessary for them, with a definite place for each implement may be derived, and a plan drawn with the exact knowledge of what it will hold.

Fig. 124 shows the working drawing of one type of shed, together with a lumber bill for it. The concrete floor is optional, but is desirable in that it raises the wheels of the machines above the surrounding grade and thus prevents rust. It should be made to slope toward the front of the shed, dropping one-eighth to one-fourth of an inch to a foot.

Doors may be added to prevent snow or rain from being driven in on the machinery.

A repair room or tool house is sometimes added.

A list of the floor space required for some standard farm implements is given on page 113.

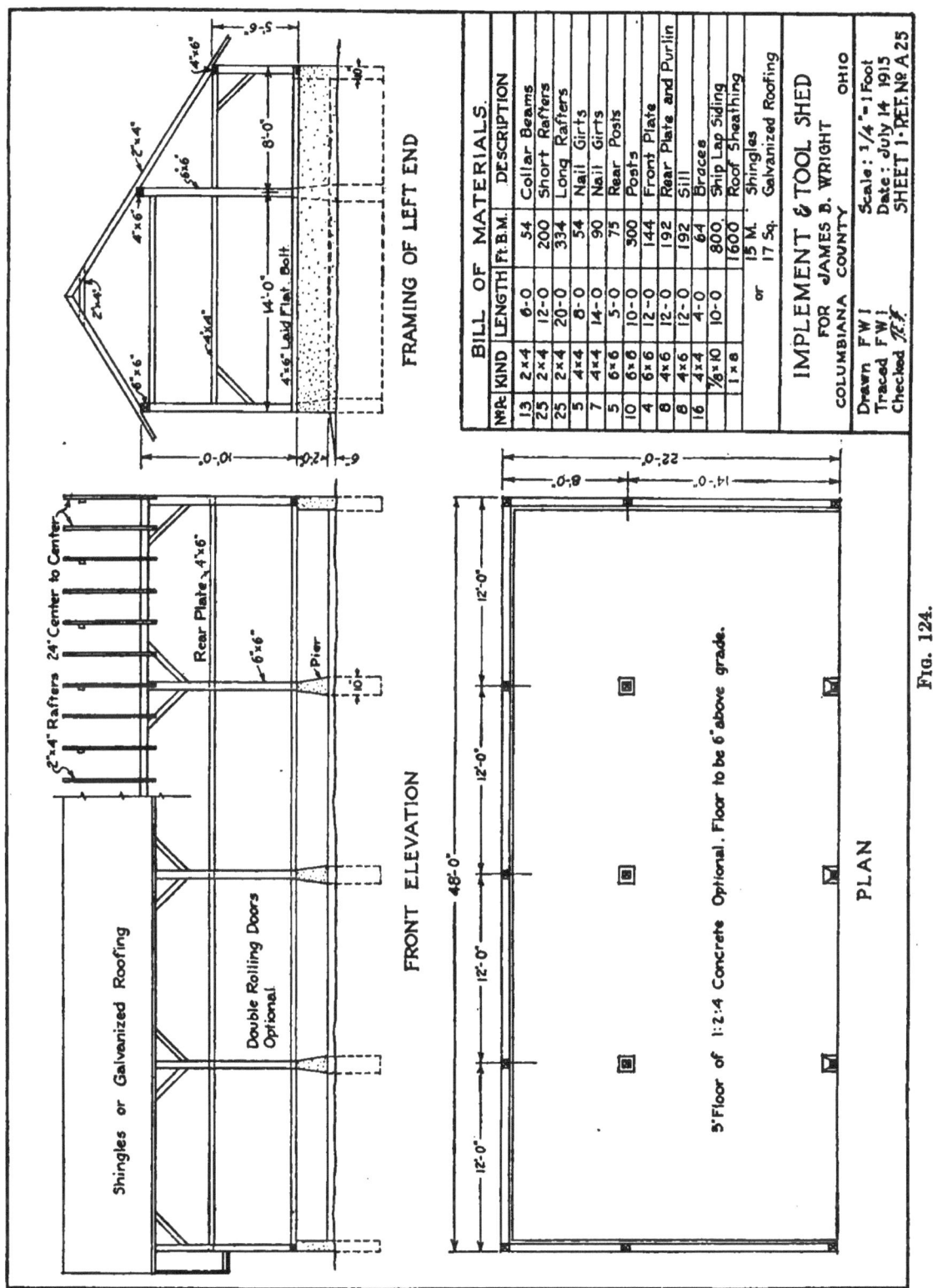

FRAMING OF LEFT END

FRONT ELEVATION

PLAN

FIG. 124.

BILL OF MATERIALS.

Nº	KIND	LENGTH	Ft. B.M.	DESCRIPTION
13	2×4	6-0	54	Collar Beams
25	2×4	12-0	200	Short Rafters
25	2×4	20-0	334	Long Rafters
5	4×4	8-0	54	Nail Girts
7	4×4	14-0	90	Nail Girts
5	6×6	5-0	75	Rear Posts
10	6×6	10-0	300	Posts
4	6×6	12-0	144	Front Plate
8	4×6	12-0	192	Rear Plate and Purlin
8	4×6	12-0	192	Sill
16	4×4	4-0	64	Braces
	⅞×10	10-0	800	Ship Lap Siding
	1×8		1600	Roof Sheathing
			15 M.	Shingles
		or	17 Sq.	Galvanized Roofing

IMPLEMENT & TOOL SHED

FOR JAMES B. WRIGHT

COLUMBIANA COUNTY   OHIO

Drawn F W I	Scale: 1/4"=1 Foot
Traced F W I	Date: July 14 1915
Checked	SHEET 1: REF. Nº A 25

Shingles or Galvanized Roofing

2×4" Rafters 24" Center to Center

Rear Plate, 4×6"

6×6"

Pier

Double Rolling Doors Optional

5" Floor of 1:2:4 Concrete Optional. Floor to be 6" above grade.

4×6"

4×4"

6×6"

2×4"

4×6" Lag Flat Bolt

CORN CRIBS

A corn crib should be constructed so as to allow the air to circulate freely through and one-half cubic feet should be allowed for a bushel of ear corn.

Rat, mouse, and bird proof cribs are easily constructed. A concrete floor will practi-

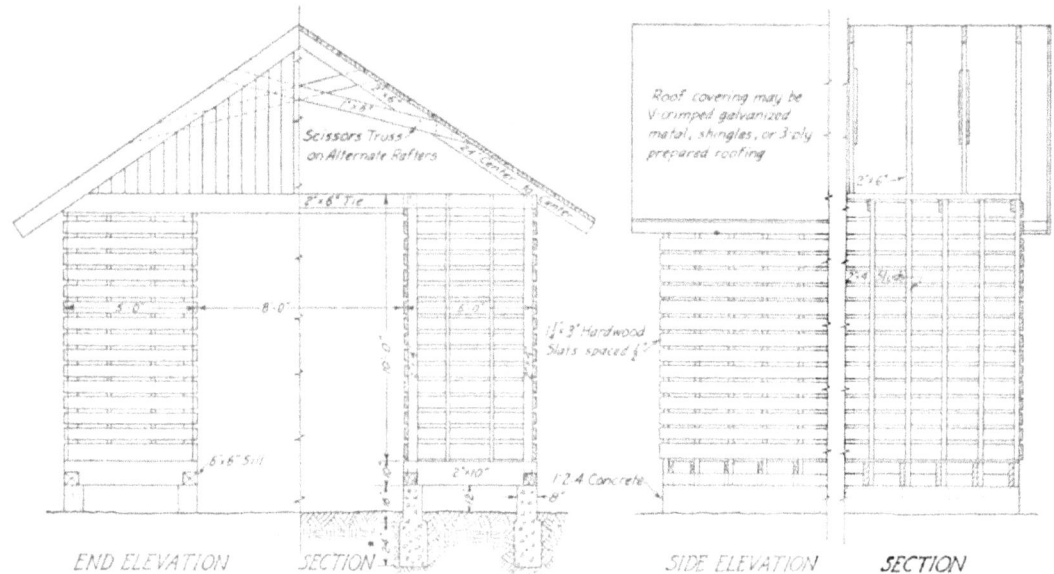

FIG. 125.—Double corn crib, stright sides, wood floor.

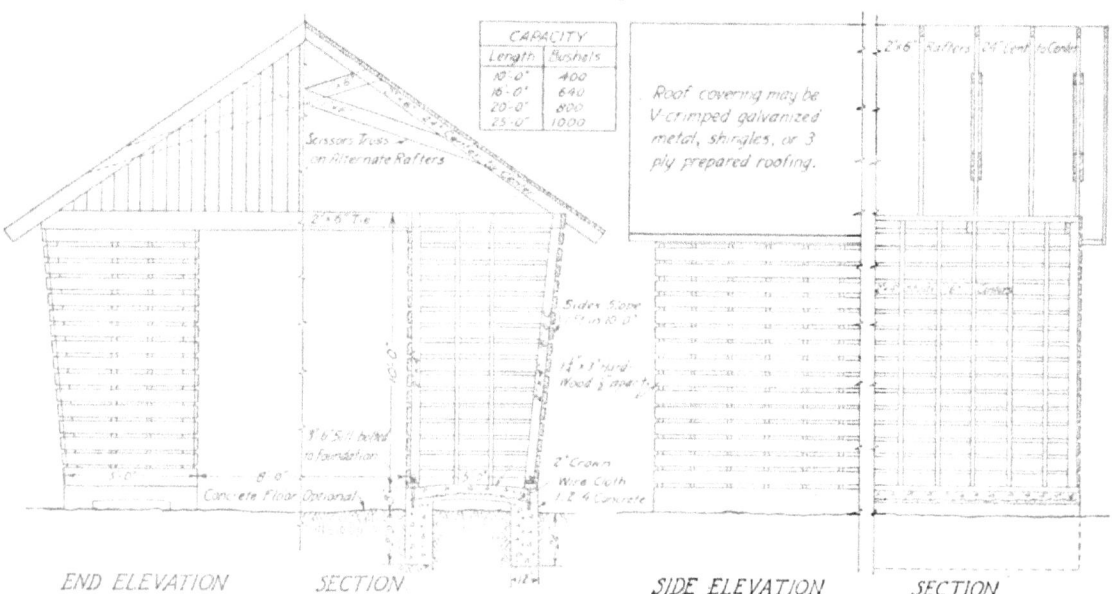

FIG. 126.—Double corn crib, flaring sides, concrete floor.

the grain. In regions where it is very moist, as in Ohio, cribs should not be built more than five feet in width. In dry localities they may be eight to ten feet wide. Two cally eliminate the two former pests. Tight construction about the eaves will prevent annoyance from the birds. One-fourth inch wire mesh screen placed under the floor slats of wood floors, and between slats and studs on the walls, will effectually stop rodents and birds.

Figs. 125 and 126 show suggested construction for two types of double crib. In addition to shelter and stability, the double crib

ings the use of the "limiting break" line is illustrated.

GRANARIES

The important consideration in the construction of granaries and bins for loose

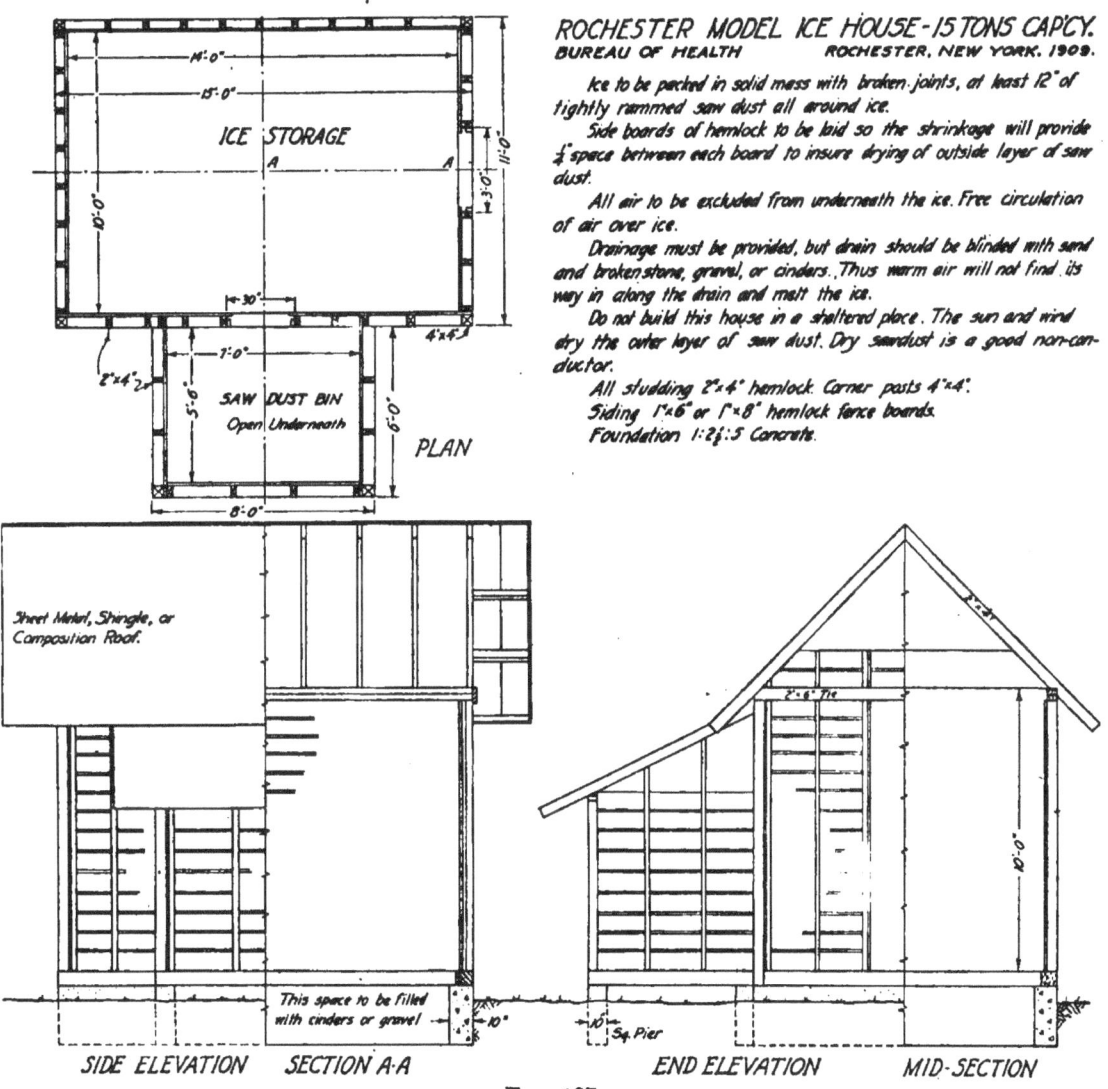

ROCHESTER MODEL ICE HOUSE -15 TONS CAP'CY.
BUREAU OF HEALTH   ROCHESTER, NEW YORK. 1909.

Ice to be packed in solid mass with broken joints, at least 12" of tightly rammed saw dust all around ice.

Side boards of hemlock to be laid so the shrinkage will provide ¼ space between each board to insure drying of outside layer of saw dust.

All air to be excluded from underneath the ice. Free circulation of air over ice.

Drainage must be provided, but drain should be blinded with sand and broken stone, gravel, or cinders. Thus warm air will not find its way in along the drain and melt the ice.

Do not build this house in a sheltered place. The sun and wind dry the outer layer of saw dust. Dry sawdust is a good non-conductor.

All studding 2"x4" hemlock. Corner posts 4"x4".
Siding 1"x6" or 1"x8" hemlock fence boards.
Foundation 1:2½:5 Concrete.

FIG. 127.

has the advantage of providing space which may be utilized for winter storage of farm implements. One of the two examples is shown with concrete floor construction and the other with a wood floor. In these draw-

grain is to be sure that the structure is designed with sufficient strength to prevent bulging of the sides and springing of the floors. In elevator and large bin construction this becomes a real engineering problem, but in any case the weight of the maximum grain contents should be figured and the structure designed to suit. The numerous recorded accidents from failure of elevated bin floors serve as warnings. Concrete, properly reinforced, is the ideal material for granaries. Concrete floors laid directly on earth should be well underdrained and damp-proofed. In any granary construction rat proofing is essential. Incidentally, rats and mice will not gnaw hemlock.

The growing tendency to build community elevators is lessening the number of farm granaries.

Ice Houses

Ice storage, formerly regarded as a luxury, is, for dairy farms at least, a necessity. The proper cooling of milk and the storage

(3) Ventilation of space above ice.

The elaborate and expensive insulated walls, with many thicknesses of different materials, are being replaced by simpler construction, the limit of simplicity being reached in the Rochester ice house illustrated in Fig. 127, which is being built by many farmers, and shows remarkable results in efficiency. In it the entire insulation is the layer of saw dust, which, when the building is properly located in an exposed position, is kept dry by the free air circulation through the openings between the boards, and dry saw dust is one of the most effective insulating materials known. The requirements for the successful construction and packing of this house are given on the drawing.

In designing a building of specified capacity, figure on the basis that a cubic foot

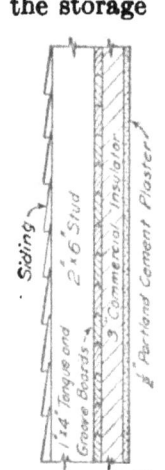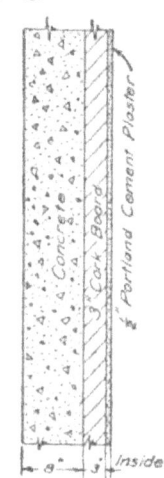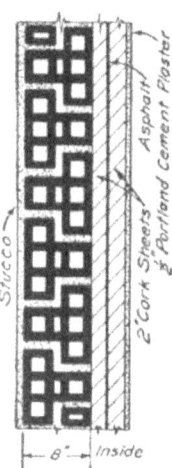

Fig. 128.—Insulated walls for ice-house construction.

of dairy products demands the use of ice in summer. The average dairy requires a storage of from one-half to one and one-half tons of ice per cow.

The requirements for ice house construction are simple but they must be carried out very carefully.

(1) There must be ample drainage, and the drainage so arranged as to exclude air.

(2) Careful insulation of walls.

of ice weighs about fifty-seven pounds, or roughly one ton of ice will require forty cubic feet, including packing. Twelve inches should be allowed on top, bottom and sides for saw dust. Several approved types of insulated walls are shown in Fig. 128.

Garages

A simple problem in building construction may be made in the design of a garage.

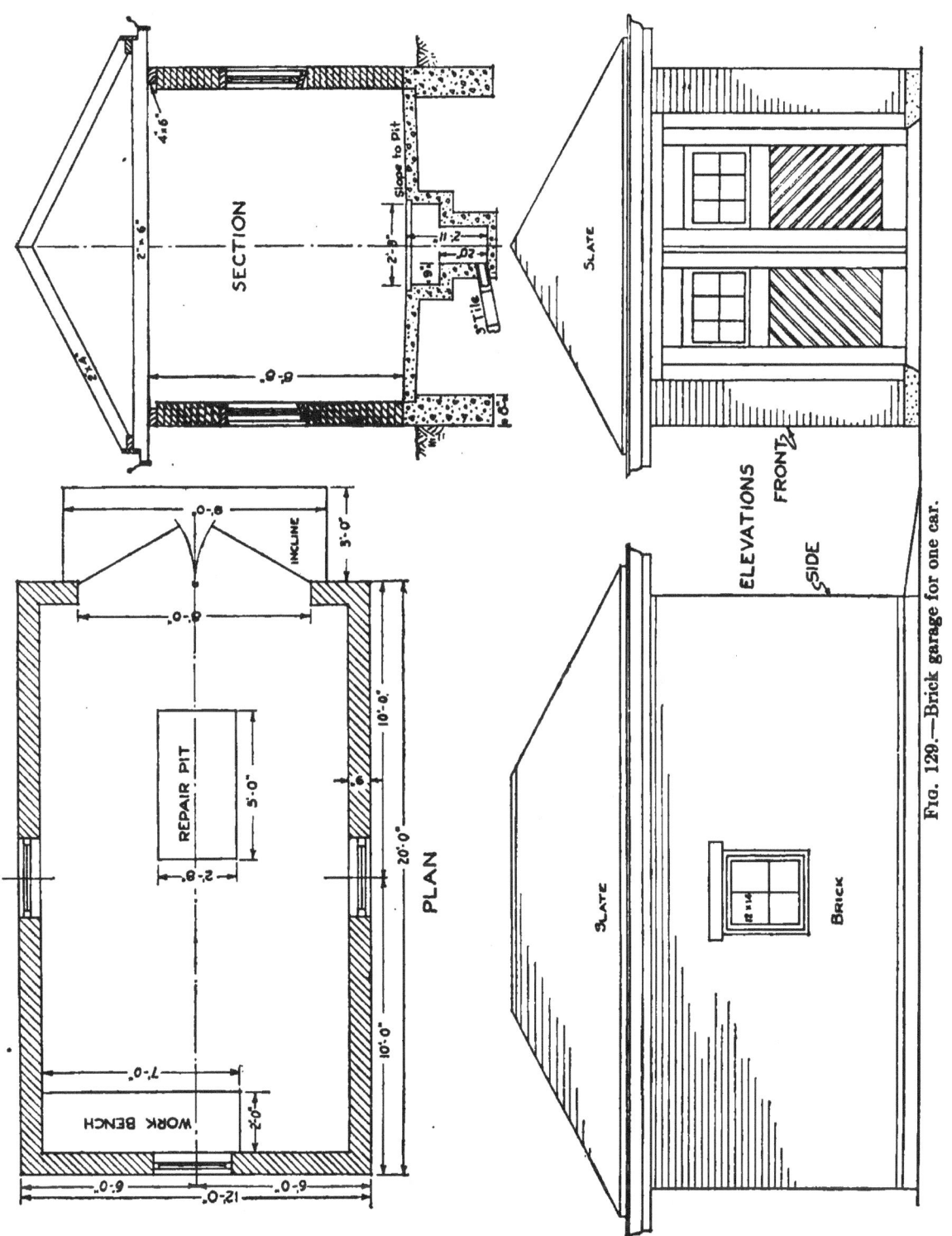

FIG. 129.—Brick garage for one car.

It may be built of wood, brick, stucco or even galvanized iron. The size of machine or machines to be housed is of course the first consideration. Provision for a work bench or shelf, the method of gasoline storage, the desirability of a repair pit, convenience of entrance and exit, conformity of the exterior to the surrounding architecture, materials for walls and roof, provision for water and drainage, best location for windows, and method of artificial lighting will all be decided in the preliminary study of the problem.

Fig. 129 shows a garage for a small car, built of brick, with a hip roof. The symbol for brick in elevation to small scale is used on the drawing.

SMOKE HOUSES

The smoke house, formerly found on every farm, is not so common a farm structure at the present day, but the problem of its design may be used, as many still prefer to smoke their own meats. It may be built of frame, brick, concrete, concrete blocks, or hollow tile, although the frame smoke house has an undesirable fire risk.

A small building, five feet square, has a capacity for eight hams. The meat should be hung at least five feet above the fire.

For an average farm, a brick house 6' by 8', having a grate arranged for burning cobs or chips, and with vents near the roof, is a common type. A safeguard in the form of an arch or sheet of metal above the fire to prevent loss of meat from blazing fat or unexpected flares, is desirable. A suspended iron rack on which the meat is laid is an improvement over the usual sticks and strings. The alternative of sharp steel hooks for suspending the meat will eliminate the unsightly holes usually made when stringing hams. The moral effect of a good strong lock should not be overlooked.

THE DAIRY HOUSE

The dairy house is an adjunct to the dairy barn, and in its simplest form is used for the temporary storage of milk. It is sometimes built in combination with the ice house; but probably oftener the cooling of the milk depends upon a water supply from a spring or well, flowing constantly through a tank in which the milk cans are set. It must be of a type that will protect against heat, dust and flies, and necessarily of simple construction so that it may easily be kept clean and sanitary. Concrete or concrete blocks, hollow tile, and brick are particularly well adapted for dairy house construction, but the ice-house type of insulated frame wall is satisfactory. Concrete floors and tanks are sanitary, permanent and easily constructed.

A very common type for keeping the night milking until the following morning, is a house eight by ten feet inside, with a tank extending across one side, opposite the door, and provided with a window, and a ventilator in the roof.

A dairy house in which an aerator, cooler, separator, and churn are to be used, becomes a problem in design, in the arranging of the equipment in the most convenient and economical form, and planning a building to suit.

THE SILO

Much has been written in recent years on the silo and its desirability, some has even been written against it, and the average farmer is well informed as to its advantages and limitations. Regarding the silo from the point of view of design and drawing it seems unnecessary to include the details of construction, either of the forms for building concrete silos, or the method of construction of wooden ones, as in the former case forms may be secured from those who make a specialty of building and renting them, much more cheaply than they can be made for use on one job, and stave, tile, and galvanized iron silos are purchased ready for erection.

In connection with a barn problem our interest is in the required capacity, and our drawing simply circles of the required diameter on the plan, and the exterior outline on the elevation.

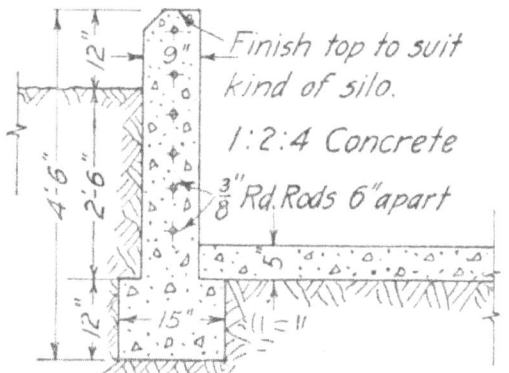

FIG. 130.—Silo foundation of concrete.

The thickness of the wall, as indicated on the plan, will depend upon the material and size. Concrete block and vitrified tile varies from 4 to 8 inches, monolithic concrete 6 inches, stave silos 2 inches, lath and plaster, or Gurler type 6″, and galvanized iron a single black line.

Fig. 130 shows a section of concrete foundation adaptable for silos of any type.

Silo tables of use in this connection are found on page 115.

THE MANURE PIT

The value of the covered manure pit for the preservation and conservation of stable manure is becoming more and more appreciated. Manure pits are simply large tanks or vats to which the manure is taken and stored until finally used on the fields. The pit should be built not less than 100 feet from the dairy stable, although most dairy requirements permit a minimum of fifty feet, and should be so arranged as to be easily accessible for wagon or spreader.

It is desirable that it be roofed, and if possible screened against flies.

Fig. 132 shows a manure pit designed for twenty head of stock. The small appended table gives the size and capacity required for herds of various sizes.

THE SEPTIC TANK

The introduction of water supply, and laundry and bath conveniences in country homes has made the provision for suitable and sanitary means for taking care of house wastes and sewage, a necessity. The open cesspool is at best a temporary affair, unsatisfactory and dangerous to health. Scientific study of bacteria and bacterial action on sewage has resulted in the recommendation of the use of the septic tank, a simple underground receptacle with inlet and outlet, in which the sewage is purified by natural bacterial action.

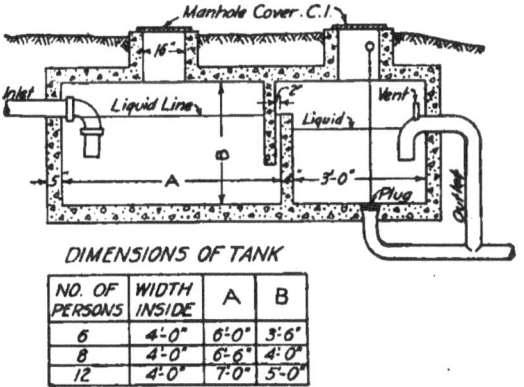

DIMENSIONS OF TANK

NO. OF PERSONS	WIDTH INSIDE	A	B
6	4′-0″	6′-0″	3′-6″
8	4′-0″	6′-6″	4′-0″
12	4′-0″	7′-0″	5′-0″

FIG. 131.—Double chambered septic tank.

The type shown in Fig. 131 is known as the double chambered septic tank, and gives better results than the single tank. Patented improvements are sometimes added.

A table of sizes is appended.

FENCES, PADDOCKS, PENS AND GATES

The fence to be effective must be strong, durable, tight, and high enough for its purpose. When a fence is made of posts supporting wire, the corner posts must be set

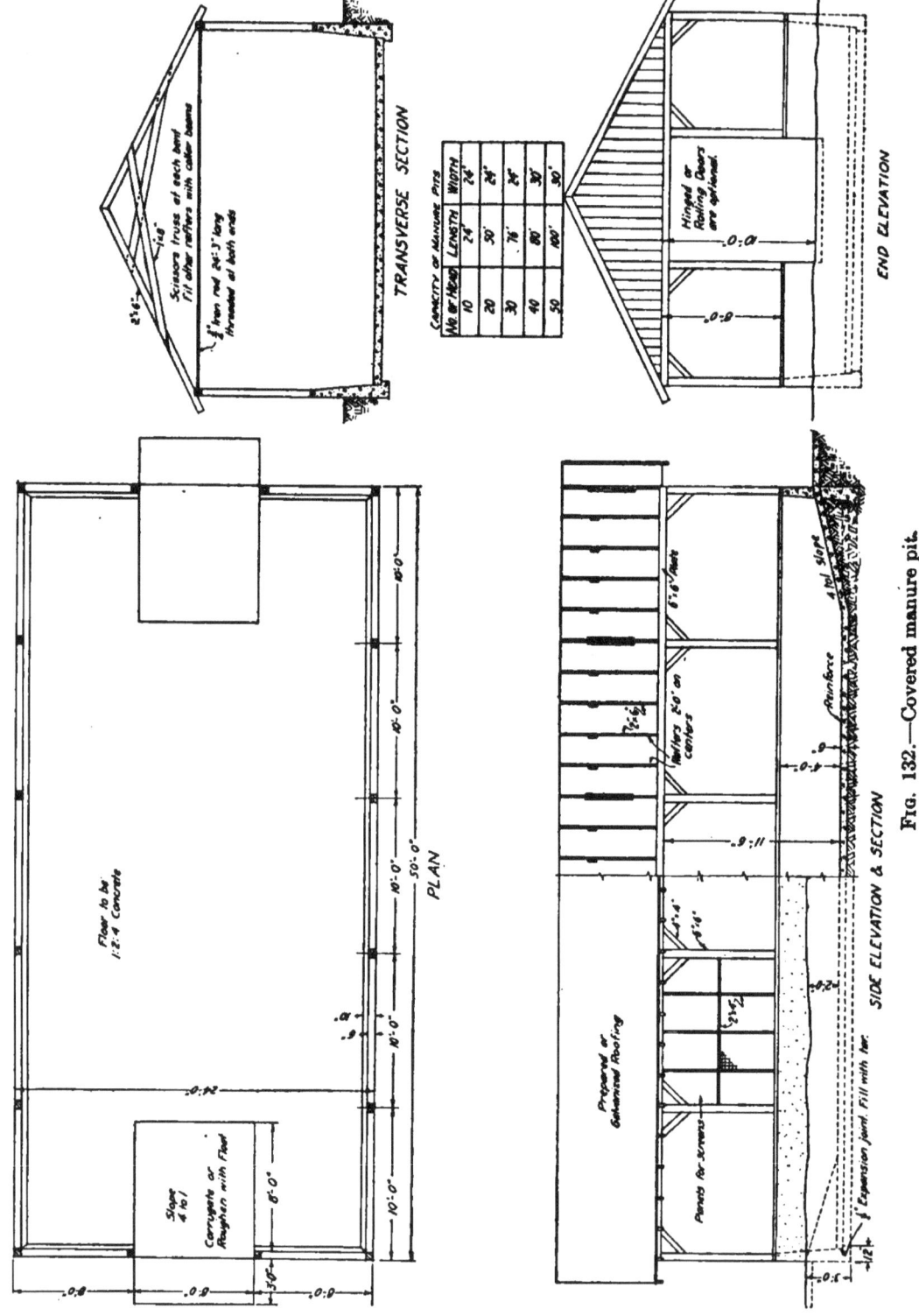

Fig. 132.—Covered manure pit.

deeply and be firmly anchored. All corners or ends must be thoroughly braced. The average wire fence is 48″ high with posts 16 feet apart and in the ground from 18″ to 3 feet, depending upon the nature of the soil. Corner and brace posts are set 4 feet deep.

Paddocks are designed for close confinement of animals, and are usually constructed

are common heights, and the sections may be of any length convenient for handling.

Pens are used for smaller enclosures than paddocks, intended for a single animal or a few animals. The construction will vary according to the animals to be confined.

Gates, as a rule should be made stronger than the fence of which they form a movable section, as they receive harder service than

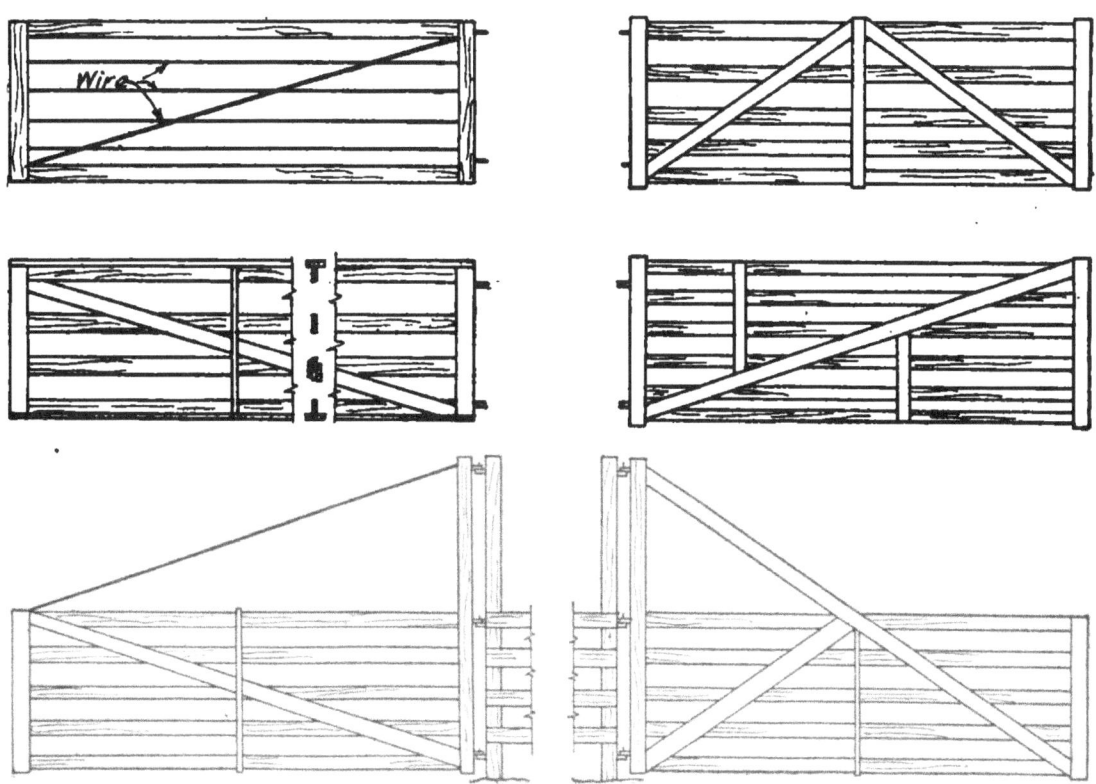

Fig. 133.—Some forms of gate construction.

of boards or planks, nailed, screwed or bolted firmly to the posts. They are made from 5 to 8 feet in height. Wire paddocks are seldom used on account of liability of injury to the animals.

Hurdles are simply movable or portable fence or gate sections constructed so as to stand alone. They are used for temporary confinement, or for aid while separating animals, and are a necessary part of the equipment of sheep barns. 36″ and 42″

the fence. A 12 foot opening will accommodate most machinery, but 14 feet is better for the passage of large loads of hay. The height should conform to the fence. Gate posts should be large enough to bear the weight of the gate without allowing it to sag. There are numerous forms of gates as built on the farm, and many types for sale in the market. Fig. 133 illustrates in diagram the lines of several different designs that may be used. The bracing in each case is designed to form a truss and prevent deformation.

The Farm House

In the design of the buildings previously considered, the principles of planning and construction involved were based on the consideration of the uses and functions of each, and the securing of the maximum of convenience, economy of space and labor, and durability. In designing the farm residence there must be added to these the items of beauty, and architectural correctness, with not only the general requirements of a farm house considered, but also the individual requirements, wishes and taste of the particular family which is to occupy the one proposed. Thus this problem becomes the most interesting as well as the most complicated of all the structures about the farm.

The house must be planned to fit both the surroundings and the family. The "lay" of the land, the direction to the road, the views toward the house and from the house, and the planting, all have a bearing on the shape and the architectural style to be determined upon. The house must look as if it belonged in its location, and should express in its appearance the individuality of the owner.

To secure all this, the service of a competent architect should be engaged, who will be able to work up the owner's ideas into a harmonious design, exterior and interior, and whose knowledge and advice on the technical points of construction will protect from mistakes. The architect is the owner's representative in dealing with the contractor. There is a temptation sometimes to accept a contractor's offer to furnish plans free, but the fee of a good architect is more than saved in his protection of the owner, to say nothing of the satisfaction of having a pleasing house.

A greater mistake is to buy a set of ready-made plans made with no reference as to where the house is to be placed or who is to occupy it.

Our discussion of the farm house is intended to assist the prospective owner in setting down his ideas, so as to present workable sketch plans to the architect, as a basis for the final plans and specifications.

Planning is simply a process of reasoning, and the thought of the needs of the family, collectively and individually, gives the reason for everything put in the plan; and success in it depends upon one's knowledge of these needs and the ability to correlate and adapt them into a well-balanced design. Where interests conflict, the advantages and disadvantages are weighed, and sometimes a desired feature must be omitted, because of inability to adapt it.

The first floor plan is the first to be made. A list of the rooms desired is written, and preliminary freehand sketches tried until an apparently satisfactory arrangement is obtained. This is worked up in a 1/8" scale drawing. These are usually made on tracing paper. Part or all of the following rooms may be included in the first floor:

1. Kitchen (and pantry).
2. Dining room.
3. Living room.
4. Wash room.
5. Den, office or bed room.
6. Stair hall.
7. Screened work porch.
8. Living porch.

The *kitchen* is in many respects the most important room in the house, and much thought should be given to its arrangement. The reasoning mentioned should be applied, in making a list of the stores and supplies and planning the most convenient place for each, and in locating the table, range, cupboards, pantry, and sink, with thought of the purpose of each, mentally answering the question, "why is this the best place, will any steps be saved?" Women walk

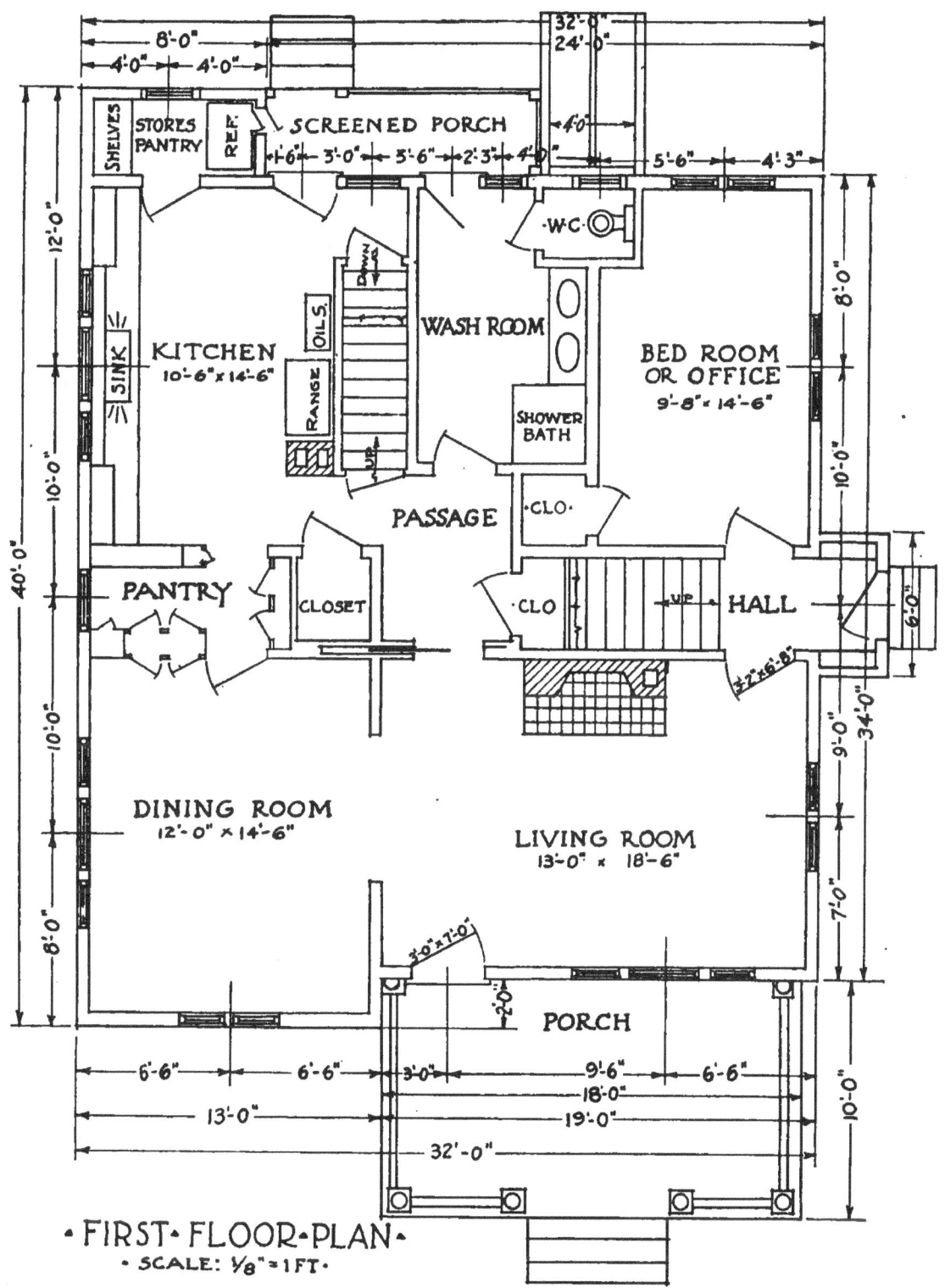

· FIRST · FLOOR · PLAN ·
· SCALE: 1/8" = 1 FT ·

FIG. 134.—First floor plan of farm house.

miles of useless steps because the men who planned their kitchens did not *think!*

The kitchen should have cross-ventilation, be well lighted and sanitary, not too large, and should open to a rear porch.

The kitchen score card given on page 117 has been used in checking the desirable points in existing kitchens, on a scale of 100.

In nearly all city and suburban houses, the kitchen is connected to the dining room through a serving pantry, which provides an air lock, keeping out odors and heat, but many in building farm houses prefer a direct double swinging door.

The *dining room* should preferably face east, should be light and cheerful, and rectangular rather than square in shape. The size can be figured accurately from the number to be accommodated, by knowing the size of a dining table and the space needed for each person. Group windows are more effective and pleasing than single windows.

The *living room* is the largest room in the house. It should have a south and west exposure, be convenient of access, have quiet continuous lines, and if possible a fire-place. The sentiment of the home centers about the fire-place.

The *wash room* is an important consideration that should not be omitted. It should open to the rear, and have ample lavatory facilities for men's convenience in coming from the field. Provision for plenty of coat hooks should not be forgotten.

A first floor *bed room* is often desired, sometimes for continuous use and sometimes for emergency use in case of illness. In the latter case a room may be designed as an office or den, and used as an emergency bed room. It should be somewhat isolated in location, and if planned for an *office* should be easily accessible from the outside.

The amount of space given to *stairs* and stair hall depends upon the economy necessary. In large houses the main stair-way is made a feature of the house, and rear stairs are always provided as well. In small houses one stairway may serve all purposes. Stairs should never be less than 3 feet wide.

In planning the stairway the number of risers should be figured, and the rule may be used that the sum of the rise and tread should be $17\frac{1}{2}$ inches. Thus if the rise is $7\frac{1}{2}$ inches the tread would be 10 inches, and on the plan the lines representing the risers would be drawn 10 inches apart. The entire run is never shown on one plan, but is broken to show what is under.

Good comfortable *porches* should be regarded as a necessity in farm house planning. A working porch off the kitchen should be screened, and a screened living porch, conveniently arranged for summer dining is a desirable consideration.

In a large house other rooms, such as a library, music room, reception room, helps' dining and sitting rooms, etc., may be added, according to the owners' desires and financial ability.

It will be noticed that the isolated and generally unused "parlor" has been omitted from the list of first floor rooms.

The *second floor* should contain

1. Hall, on which all rooms should open.
2. Sleeping rooms, with ample closet space in each.
3. Bath.
4. Linen closet.
5. Rooms for help.
6. Sewing room.
7. Sleeping porch.

In planning the second floor it is not necessary that all walls run through from the first floor partitions. The outside walls, stair well and chimneys are traced from the first floor, and the space cut up to the best advantage. No room should be made to serve as a passageway to another room.

Bed rooms should have at least two windows, with cross-ventilation if possible. The first consideration in a bed room is a well-

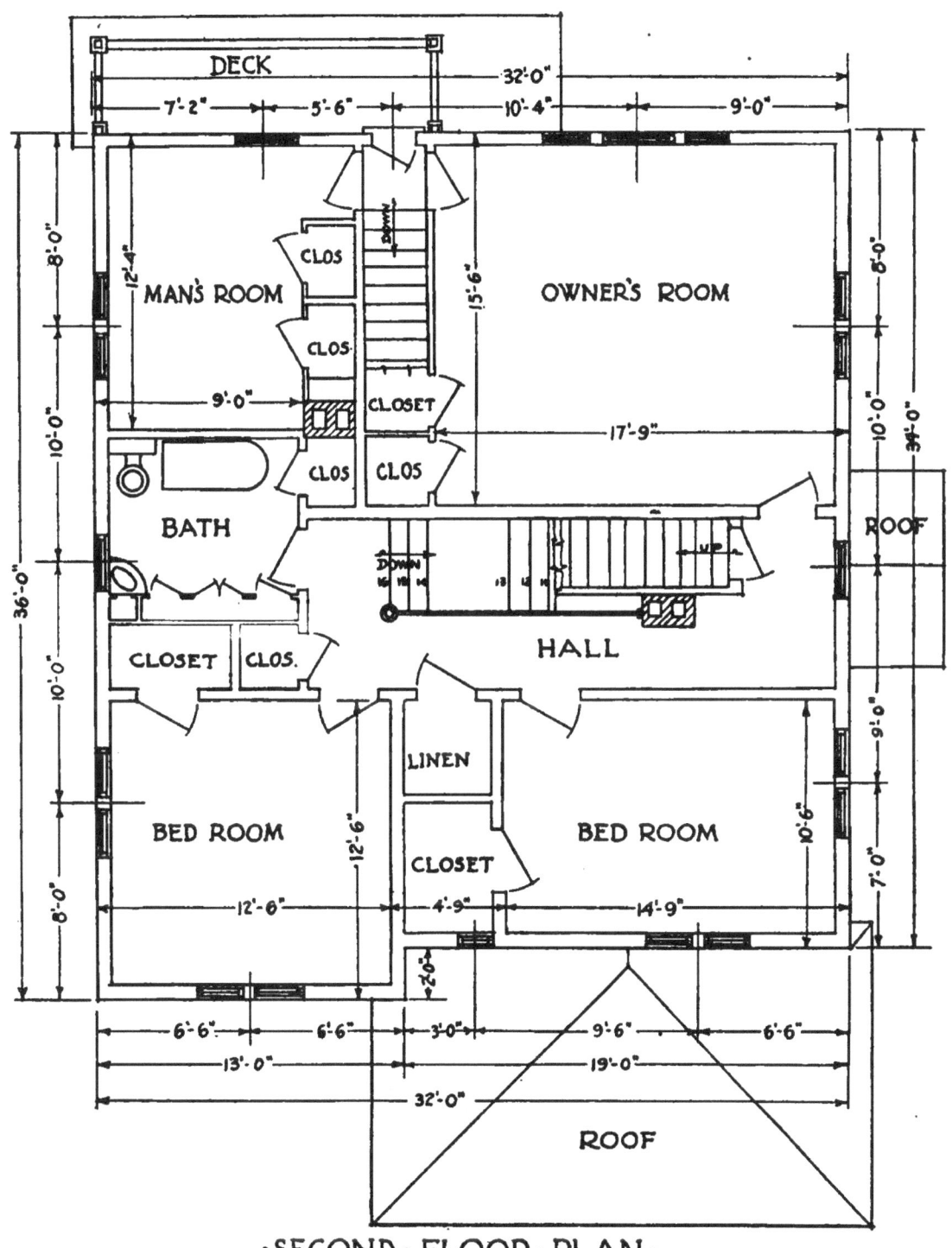

·SECOND·FLOOR·PLAN·

· SCALE: ⅛"=1 FT ·

FIG. 135.—Second floor plan of farm house.

planned space for the bed, the second is ample closet room. External angles should be avoided, that is, closets should not break into the rectangle of the room, nor should they be cut off across corners.

The *bath room* should be located so as to make the plumbing simple and economical. It need not be larger than 6' × 9'. All the fixtures should be drawn, to show their location. For a farm where much help is employed it is very desirable to have more than one bath room.

A large *linen closet* opening to the hall is very desirable. A clothes chute to the laundry may be provided.

Rooms for help are best arranged to be reached from a rear stairway and isolated from the rest of the second floor.

A small *sewing room* is a convenient addition.

A *sleeping porch* is becoming a popular addition to every house plan. It should be tightly screened, and canvas curtains provided for inclement weather. These may be fastened with carriage buttons, or mounted on rollers. The porch may open from the upstairs hall, or a private sleeping porch may open to a bed room or dressing room.

Attic stairs should always start from the hall, and not from any room. Generally on house plans the attic plan and roof plan are combined in one figure.

The basement should extend under the entire house, with concrete floor and drains. It should have an outside grade entrance as well as the inside cellar stairs. If a furnace is used, the furnace and fuel room should be separate from the storage room.

A basement laundry and drying room, with water and flue connections should be arranged. A clothes chute opening into the laundry is a convenience.

Modern farm houses often have the lighting and water supply systems in the basement.

The probable cost of a proposed house may be estimated roughly by cubing it, as explained under "estimating" on page 118.

If after drawing the preliminary sketch plans the estimate runs higher than the owner wishes to go, the process of "cutting" must be resorted to, by simplifying construction, changing materials, cutting down sizes of rooms, or omitting some. The mistake should not be made of having many small rooms. It is better to have fewer and larger ones. Radical cutting may mean the entire revision or discarding of the first plan.

Figs. 134, 135 and 136 show the basement, first floor and second floor plans of a farm house, embodying some desirable features. These plans illustrate the use of the symbols and the appearance of the usual set of plans.

Problems.

The following problems are to have complete working plans drawn, with dimensions, title and bills of material. The sizes of buildings are not given, but are to be determined from the requirements stated, figuring the space required for animals and storage and planning its practical and economical distribution. Some of the problems are suggestive rather than definite, and the data is to be assumed by the student, or assigned by the instructor.

Before starting a problem read the discussion carefully, look up other reference reading, and supplement the information found from your own practical knowledge and experience.

PROBLEMS.

1. Design and make working drawings for the forms of a concrete watering trough to hold ten barrels of water when filled within two inches of the top. Provide concrete paving six feet wide on three sides of the tank. Select suitable scale and follow directions for working given on page 38.

2. Draw an "A" type of hog cot, with 6'-0" × 8'-0" floor. Make end view first and project side view from it. If necessary, use an auxiliary projection.

3. Design a machine shed for the following implements: Two wagons, one grain binder, one corn binder, one grain drill, one mower, one side delivery hay rake, one hay tedder, one common rake, one hay loader, one disc harrow, two peg tooth harrows, one gang plow, one sulky plow, one ensilage cutter,

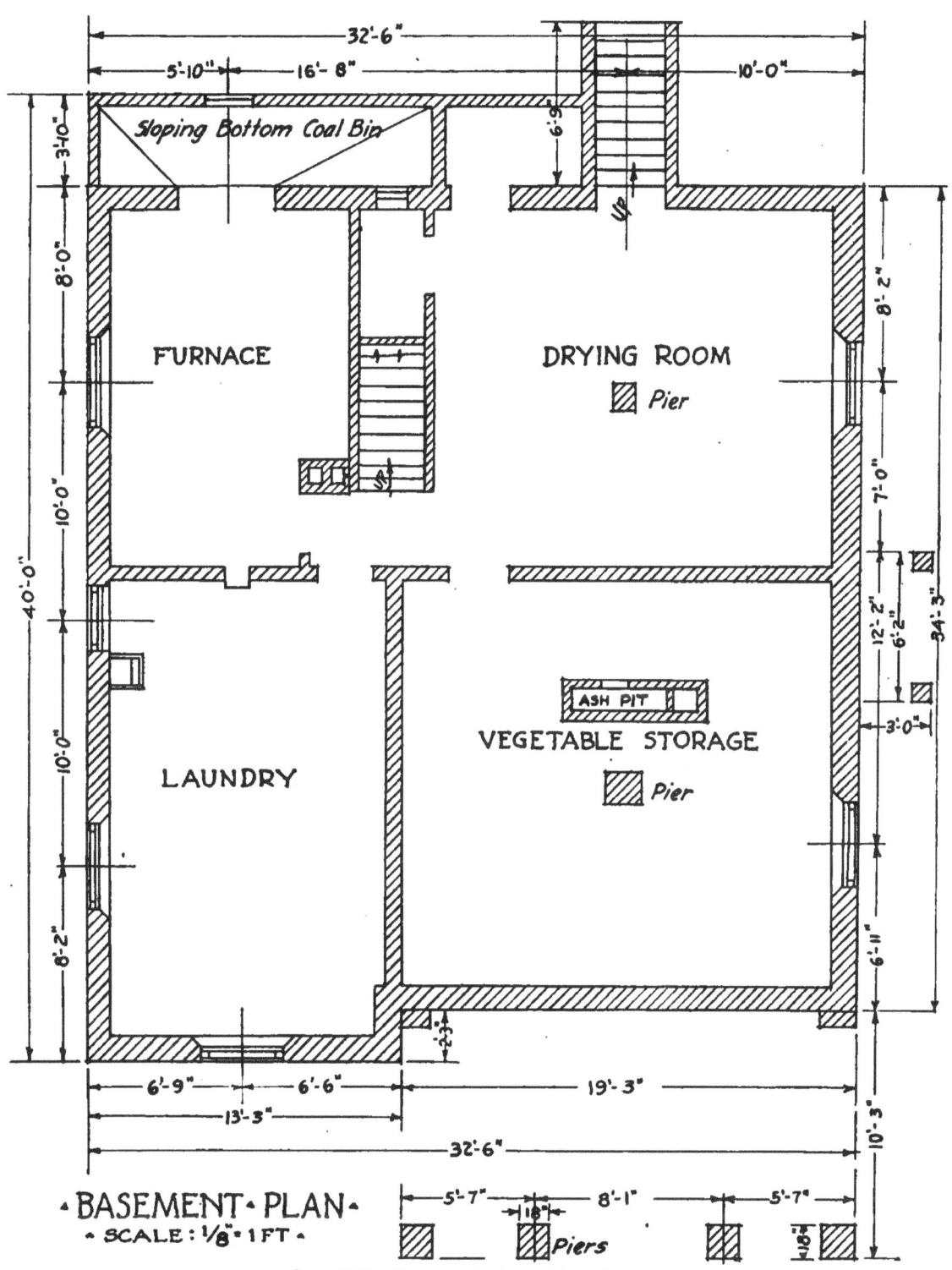

FIG. 136.—Basement plan of farm house.

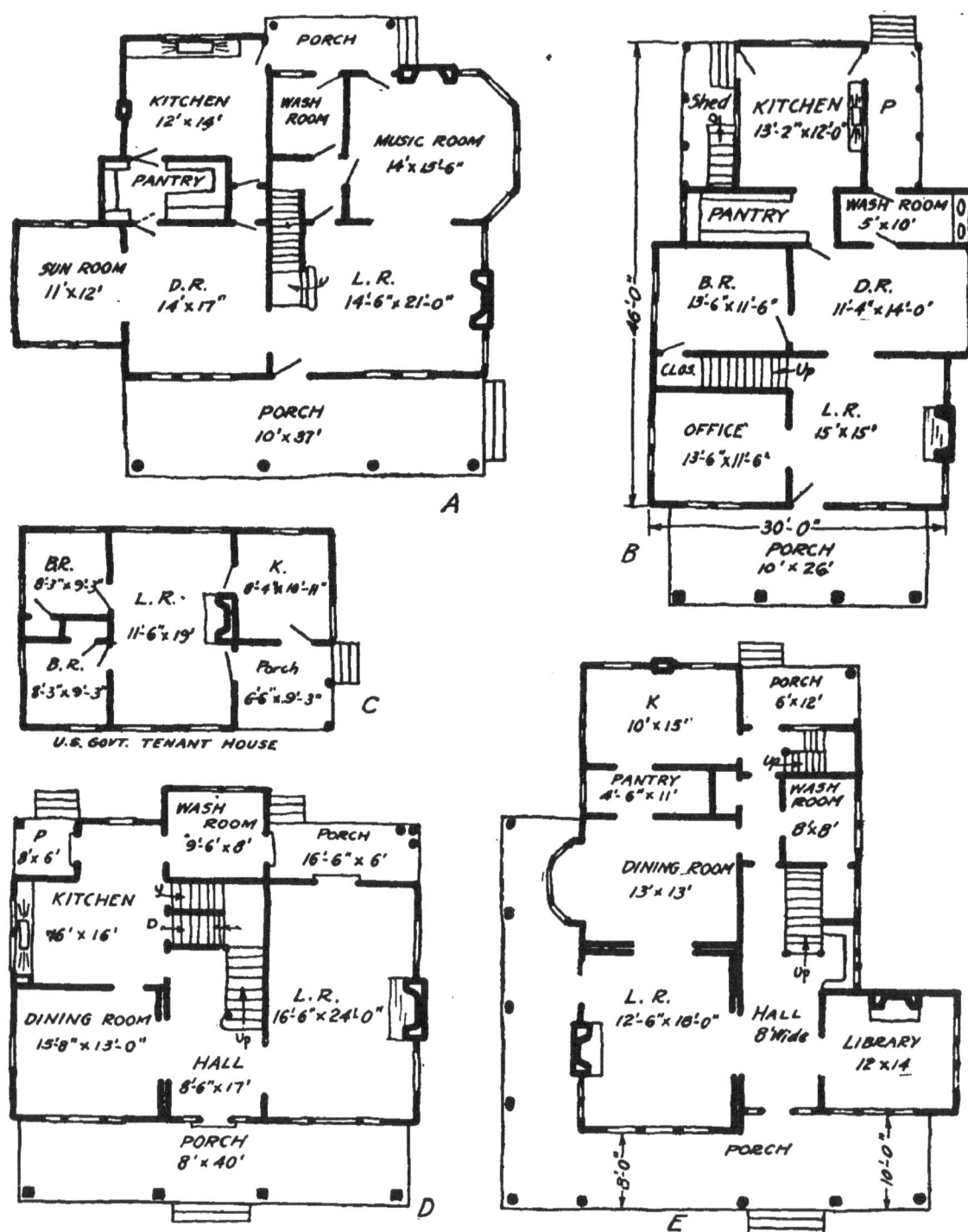

Fig. 137.—Sketch plans of farm houses.

two 2-row cultivators, one single row sulky cultivator, one walking plow, one small cultivator, one 2-row corn planter, one manure spreader, one spring wagon, and miscellaneous small tools. Add twelve feet for a tool repair room.

4. Design a machine shed for the implements in use on your farm.

5. Design a laying house for 50 hens.

6. Design a poultry house for 200 birds. This may be divided into two compartments.

7. Design a sheep shed for 200 breeding ewes. (There will be 50% more lambs than ewes in the average flock.)

8. Design a colony type house for ten brood sows. In the plan show runs and dipping vat. Provide feed room, and bedding storage.

9. Design a colony type house for twenty brood sows. Provide central storage and feed house.

10. Design a single corn crib, 500 bushels capacity.

11. Design a double corn crib, 1200 bushels capacity.

12. Design a dairy barn for 12 cows, making provision for box stalls and calf pens. Feed room to be separated from, but accessible to dairy stable.

13. Design a dairy barn for 30 cows. Compute sizes and add silos, hay storage and grain bins. To the barn attach a straw shed, the first floor of which is a covered yard, the shed capacity to be two tons of straw per animal. The shed may be arranged either to form an ell, or may be parallel to the main structure.

14. A farmer has 8 milch cows, 8 work horses, and is feeding 25 steers. Design a general purpose barn to house this stock and provide necessary feed. Allow 25 sq. ft. of floor space for each steer. (Try L or T shape in preliminary sketches.)

15. Design a rectangular general purpose barn for 6 regular horse stalls, four box stalls for brood mares, and a dairy stable for 16 cows. The dairy stable must conform to the score card on page 116. It is suggested that this be made a bank barn, with a bridge leading to the approach, so that there may be no interference with light on the bank side.

16. Design a manure pit for the barn of Problem 13.

17. Design a manure pit for the barn of Problem 15.

18. Design a garage for a ——————— machine.

19. Design a combination garage and carriage house, using stucco on 8″ hollow tile walls.

20. Design an ice house for 10 tons of ice.

21. Design an ice house for a dairy farm having 24 cows, and a dairy house provided with cold storage room and cooler.

22. Design a brick smoke house.

23. Design a septic tank for a family of 6 persons.

24. Design a root cellar of 600 bushels capacity, built four feet below ground; the top a concrete or brick arch, to allow of covering with earth. The concrete arch should be at least 6″ thick and strongly reinforced, with a span of not over 8 feet. Have the entrance on south end.

25. Design a storage barn for 75 tons of hay.

26. Design a dairy house, made of concrete blocks 8″ × 8″ × 16″, figuring sizes and spacing of openings to fit these blocks. Provide a tank 18″ deep and not less than 17″ wide. Show piping for continuous flow of water. (If tank and floor are elevated above grade it will save labor in handling cans.)

27. Design a water storage tank with a capacity of 50 barrels, having a base 6 feet square, or 6 feet in diameter, inside. The tank should have a floor

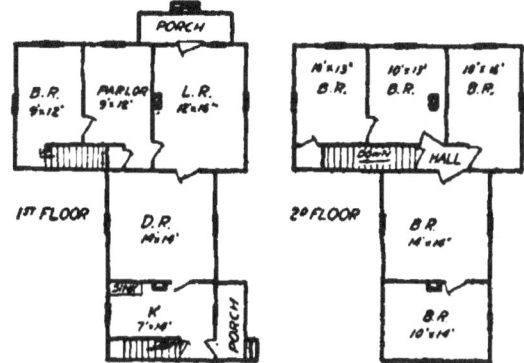

FIG. 138.—A real example of poor planning.

8″ thick, reinforced in both directions with ⅜″ rods spaced 6″ apart, and bent to extend up into the walls 2′–6″. The walls should taper (on the inside) from 10″ thick at the base to 5″ at the top, and a footing must be provided. The walls should be reinforced with ⅜″ vertical rods spaced 16″ apart, and ¼″ horizontal rods spaced 6″ apart for first 30″, 8″ apart for next 30″, and above this ⅜″ rods spaced 7″ apart. If square or rectangular tank is made the horizontal rods should be bent so as to turn corners.

28. Design a granary to hold 1200 bushels of wheat and 600 bushels of oats.

29. Make a sketch and bill of material for a paddock for horses, to enclose one acre. Have it 6 feet high, using 2″ × 6″ × 16′ plank, bolted to posts 8 feet from center to center; and provide suitable gate.

30. Design a farm house, selecting the scheme from one of the sketches of Fig. 137.

31. Design a tenant house, containing a kitchen, combined dining and living room, and two bed rooms.

32. Remodel the house shown in Fig. 138, to make it more convenient and comfortable. (This figure was taken from an actual house!)

CHAPTER V

MAPS AND TOPOGRAPHICAL DRAWING

An important consideration in both the ownership and management of a farm is to have accurate maps and plats, in ownership so that there may be no question nor dispute regarding lines or boundaries, and in management, in order to plan work, and keep a record of crop production, soil fertility, rotation of crops, etc.

A farm map gives at once a comprehensive idea of the entire farm and the relative areas, sizes and locations of the parts composing it; and imparts to the owner or manager, and incidentally to the interested visitor, a better conception or mental picture of these relations than can be gained without it. It gives a kind of birdseye view which is of great assistance to him in planning his work, in showing the most desirable locations for roads and lanes, for division of fields, and for the addition of improvements.

This chapter takes up briefly the methods and application of this branch of the language of drawing.

Instruments.

A surveyor uses a transit for turning off angles and for obtaining magnetic bearings or directions of· lines. It is an expensive instrument and, while desirable, is not a necessary investment on the farm; but a small farm level (prices of which range from ten to forty dollars) will be of constant value on every farm, for laying out drains, leveling building foundations, determining fall of streams for water power, and laying out road and fence lines. If provided with a horizontal circle, as many are, it will be found very valuable in laying out fields.

For rough determination of levels for grading or ditching, a hand level may be used. An "A" frame for this purpose is illustrated in Fig. 172.

Land measurements were formerly given in rods, and measurements made with a "Gunter's chain" of 4 rods (66 feet), divided

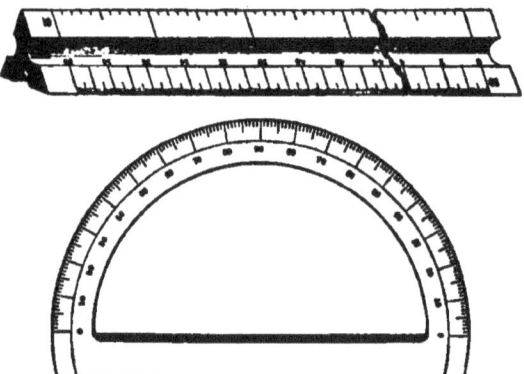

Fig. 139.—Engineers' scale and protractor.

into 100 "links" of 7.92 inches. Lengths are now given in feet and tenths of feet.

For general use on the farm a 100 ft. steel tape, graduated in feet and inches will be of constant service in laying out and measuring fields, buildings and boundaries; and for the few times when tenths of feet are needed they may be calculated from inches.

A farm office should have on hand a *protractor* for laying off angles on the map, and a 12 inch *engineer's scale*, divided into 10, 20, 30, 40, 50 and 60 parts to the inch. With it a map may be made to any desired scale, or number of feet to the inch; or if the old-style measurements are preferred, may be made to such a scale as 10 chains to the inch. The protractor and engineer's scale are illustrated in Fig. 139.

A *map* is virtually a top view of the area represented, showing the natural features and the works of man, and the imaginary lines representing divisions of authority or ownership. The relief, or relative elevations and depressions of the surface of the ground, is sometimes added, to give the complete description of an area, in which case the map is called a contour map.

The *scale* of a map varies of course with the extent of territory to be shown on a sheet of given size, but in any map the scale is relatively so small that objects are shown symbolically. The map of an ordinary farm may be drawn from 40 feet to 100 feet to the inch.

Plats.

A map plotted from a plane survey, and having the relief omitted, is called a "plat" or "land map." Under this head would come the plat of a farm survey, made from the surveyor's record, or from the description in the deed to the property.

A Farm Survey.

The plat of a farm survey should give clearly all the information necessary for the legal description of the parcel of land. It should contain:

(1) Lengths and bearings of the several sides.

(2) Acreage.

(3) Location and description of monuments found and set.

(4) Location of highways, streams, etc.

(5) Official division lines within the tract.

(6) Names of owners of abutting property.

(7) Title and north point.

(8) Certification.

Under these heads, (1) a survey to be legal must be made in "metes and bounds." The bearing of a line is its deviation from north and south. Thus a line N. 15° W. would be a line deflecting at an angle of 15 degrees to the west of the north and south line. (2) The acreage of a farm or field is calculated by well-known methods. (3) Monuments are markers set in the ground, or sometimes on the older surveys, blazed on a tree. The importance of good monuments should be emphasized, and wooden stakes should be avoided. An iron pin or pipe is sometimes used. A better marker is a gas pipe stake set in concrete in a land tile. When it is necessary to move and put back a monument, as for example in road improvements, witness stakes should be set first, with which the monument may be aligned. (4) Highways and streams should be indicated by the symbols of Fig. 142. Waterlining if attempted, should be done carefully. If a boundary runs up the middle of a stream on a farm survey, it is best to omit the waterlining. (5) Official division lines would mean township and range lines, or other legal divisions. (6) Adding the names of the owners of abutting property aids in fixing the location of the land in relation to adjacent property holders. (7) The title to a map is a concise formal statement of general information, giving name of owner, location, purpose of the map, when made and by whom, scale, date, and key to symbols. The North Point serves to orient the map, and a note should state whether the true or magnetic meridian is referred to. (8) A survey to be legal must be made by a surveyor, and be certified as correct by the County Surveyor's office. The owner should familiarize himself with the State laws of his state.

Fig. 140 is an example of a farm plat.

A Farm Office Map.

In addition to the items of the farm survey, a map for office use should contain—

(1) Location of buildings.

(2) Location of all drains and underground pipes.

(3) Division lines of all fields, fenced or unfenced.

This map, with the fields numbered and recorded in a farm book will be a valuable aid in recording crop rotation, production, state of fertility, soil tests, dates of improve-

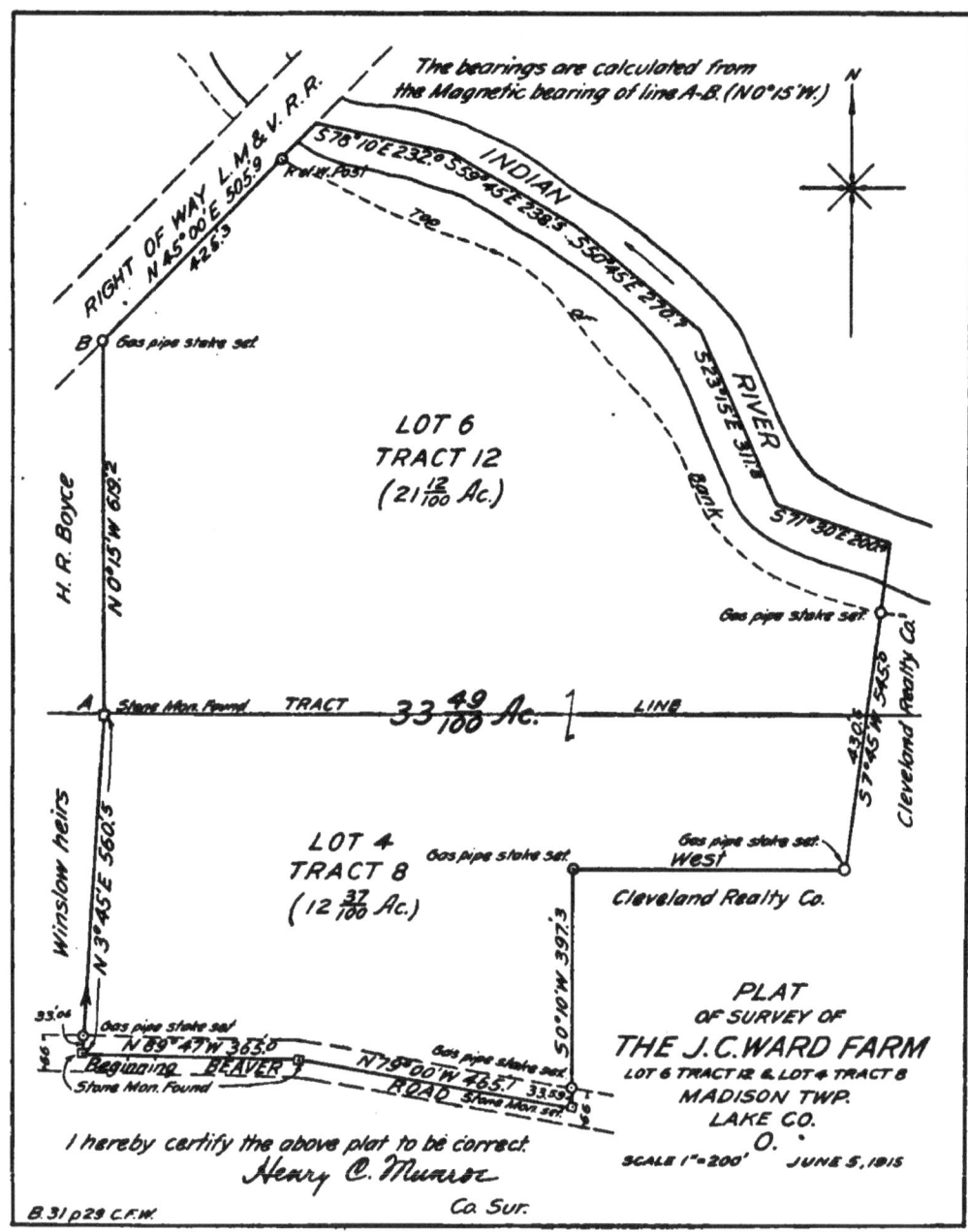

FIG. 140.—Farm plat from plane survey.

(4) Numbers on all fields.

(5) All lanes and drives.

ments, etc. It will also serve in directing farm labor, and planning work ahead, such

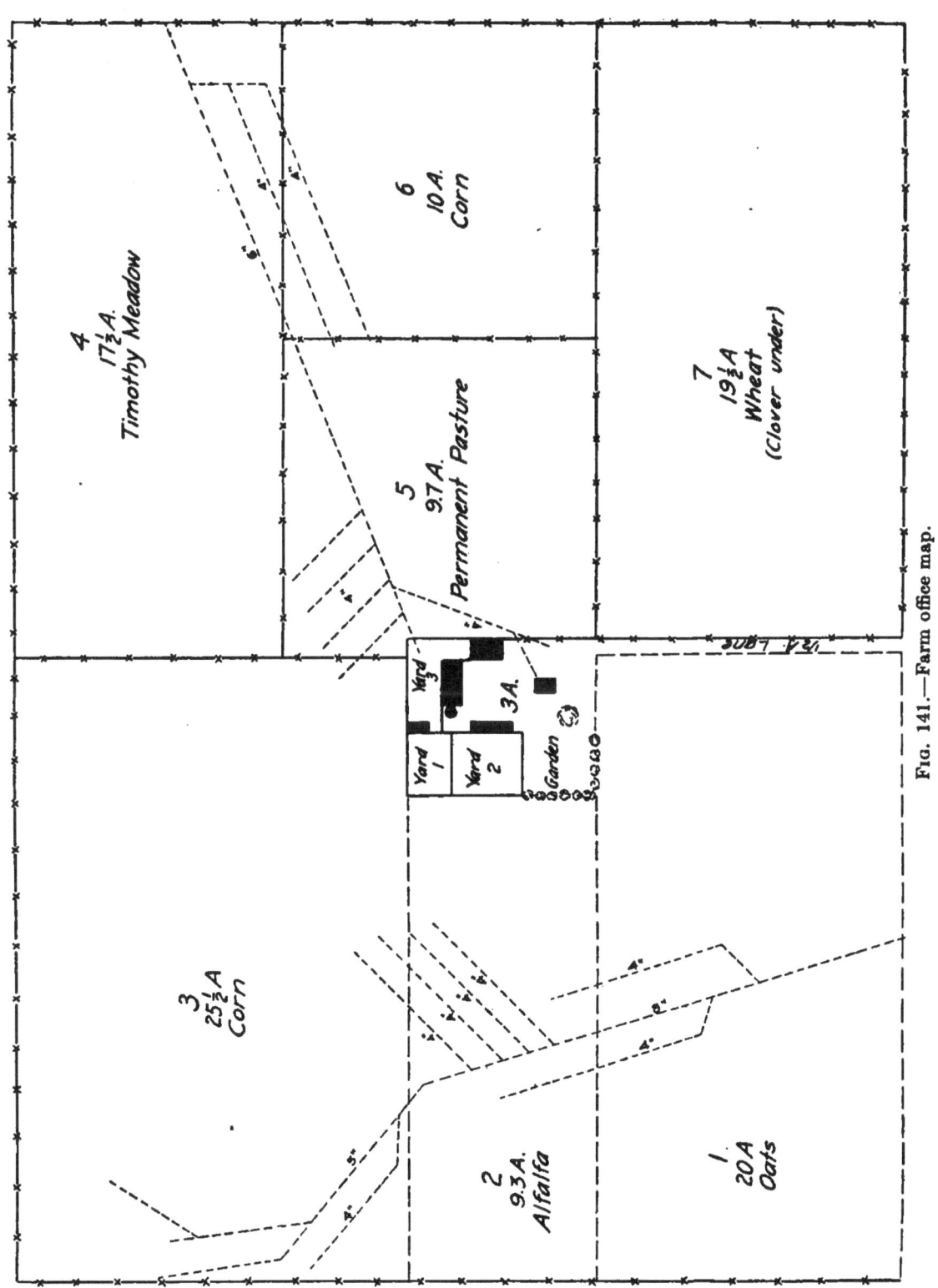

Fig. 141.—Farm office map.

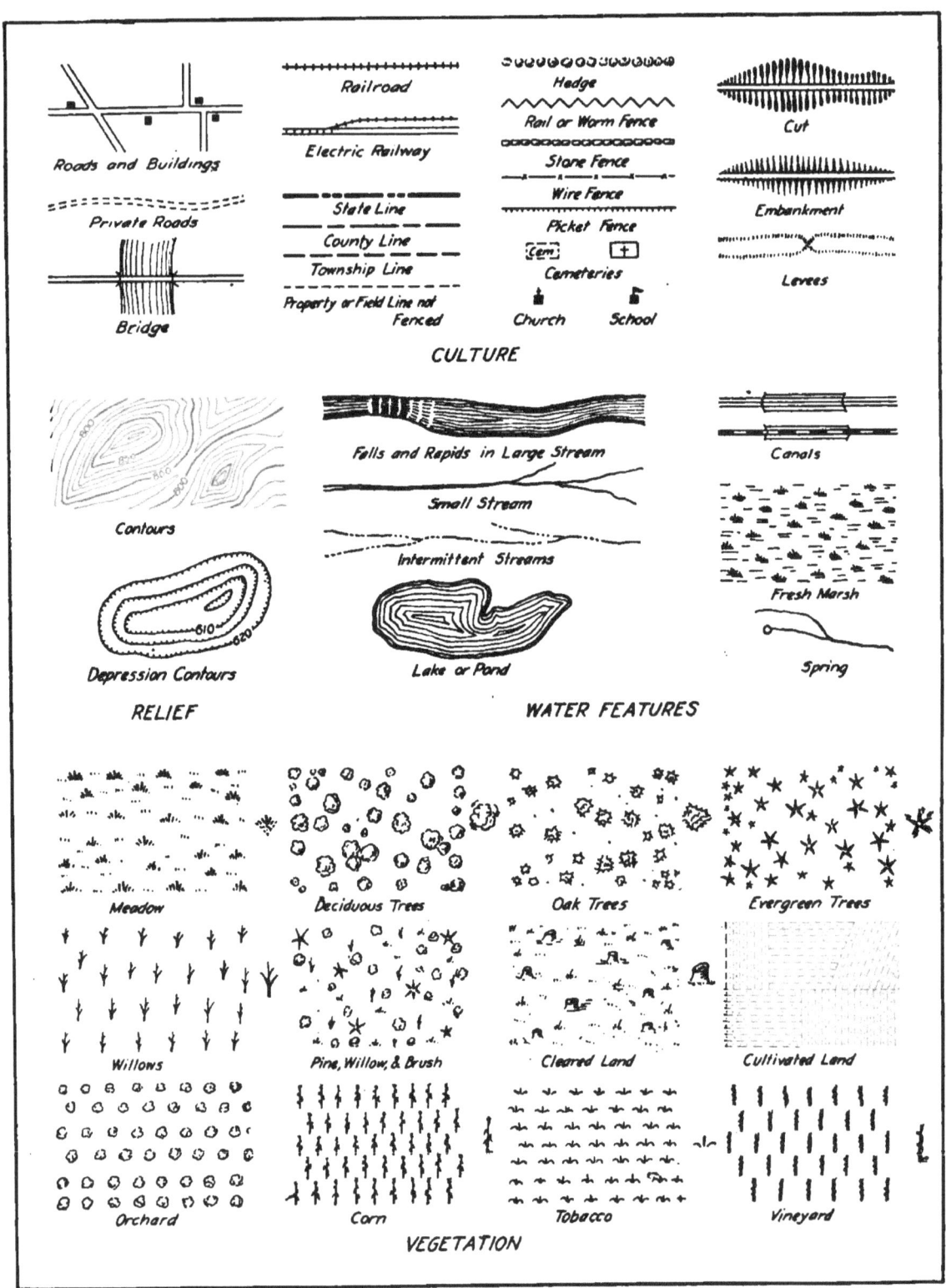

Fig. 142.—Topographic symbols.

as fence repairs, plowing, etc. Some use such a map, mounted on soft pine, or cork board, by using pins with variously colored heads in locating and recording operations.

Sometimes the contour lines are drawn on the farm map to aid still further in its use, particularly in planning drainage.

If the map is made on tracing cloth, blue prints or blue line prints may be made, and the original tracing kept up to date in regard to permanent improvements, while the

lines, the positions of natural land and water features, and the "culture" (as topographers call the works of man), but also the relative elevations and depressions, or the slopes of the hills and valleys. The different kinds of planting and vegetation are indicated by symbols, and the relief is generally shown by contour lines.

Fig. 142 is a page of standard symbols used in topographic drawing. They should be made with a fine pen and not overworked.

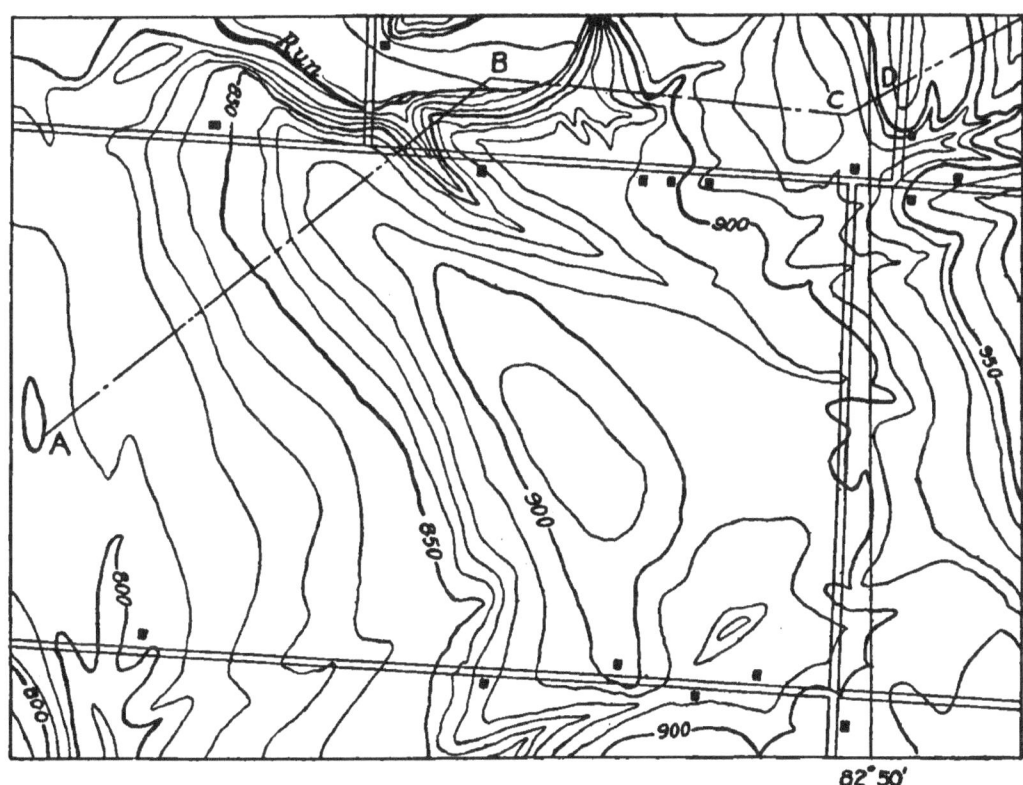

Fig. 143.—Contour map.

prints are used to work from, and filed from year to year. In extensive operating a print may be given to each tenant.

Fig. 141 shows a farm map for office use.

Topographical Drawing.

A topographic map is a more nearly complete description than a land map, in that it gives not only the imaginary division

Contours.

A contour is a line on the surface of the ground which at every point passes through the same elevation, that is, the same "level." Thus the shore line of a body of still water represents a contour. If a portion of a farm should be flooded, the edge of the water would be one possible contour. If the water went down one foot in depth, the new shore line would form a new contour, with a contour interval of one foot.

Fig. 143 illustrates a contour map, in which the contour intervals are ten feet for the light lines and fifty feet for the heavier lines. Contour lines are also shown in Fig. 5, page 4.

Quadrangle Sheets.

The U. S. Government, in cooperation with the different states, and under the direction of the Geological Survey, is mapping each state, in sections called quadrangles, on separate sheets of about 16″ × 20″, and mostly to the scale of approximately one inch to the mile. These maps show all

features in inclined letters. For plats, the letters of Fig. 42 are well adapted. Notes on contour and profile maps are generally made in Reinhardt letters; Fig. 46. On landscape maps the Roman of Fig. 47 may be used.

Profiles.

If a vertical section is taken along the line *ABCD* of Fig. 143 the view is termed a profile, as shown in Fig. 144. Here it is seen that the vertical distances are exaggerated, or plotted to a larger scale in order to show the grades to better advantage. Profiles are usually made on ruled profile paper, that known as "Plate A" paper, with 4

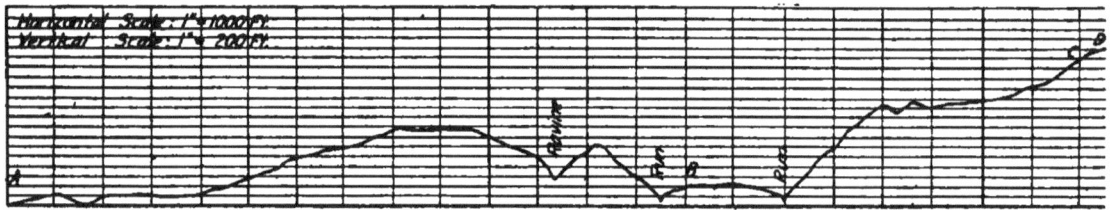

FIG. 144.—Profile from contour map.

roads, houses and other topography, together with the contours, and give the elevations of all cross roads, in feet above mean sea level. They are distributed at the nominal cost of ten cents each, and will be found of much value and interest to any land owner. To find if a particular section has been completed, write to the Director, U. S. Geological Survey, Washington, D. C.

Landscape Maps are used in the study of improvements for estates, country places, parks and additions, and for showing the artistic effect to be gained in the arrangement and planting. Some degree of embellishment is permissible, and color is sometimes used on them. Fig. 145 illustrates a landscape map of an addition.

The *lettering* done on maps should be simple and legible. Land features are usually indicated in vertical letters and water

horizontal divisions and 20 vertical divisions to the inch, being generally used. They are of value in grading drains and highways.

The gradient, or grade of a line is the percentage of its vertical rise or fall, thus a road of 7 per cent. grade would rise seven feet in a distance of 100 feet, and a drain with −0.5 per cent. grade would have a fall of 6 inches in 100 feet. Tile drains should not have less than −0.1% grade. Open ditches vary from 2 to 8 feet fall per mile.

Problems.

The following problems illustrate the type that may arise in connection with this branch of drawing, and will serve as suggestions for the student or farm owner; and will give a familiarity with the symbols, and the method of using maps.

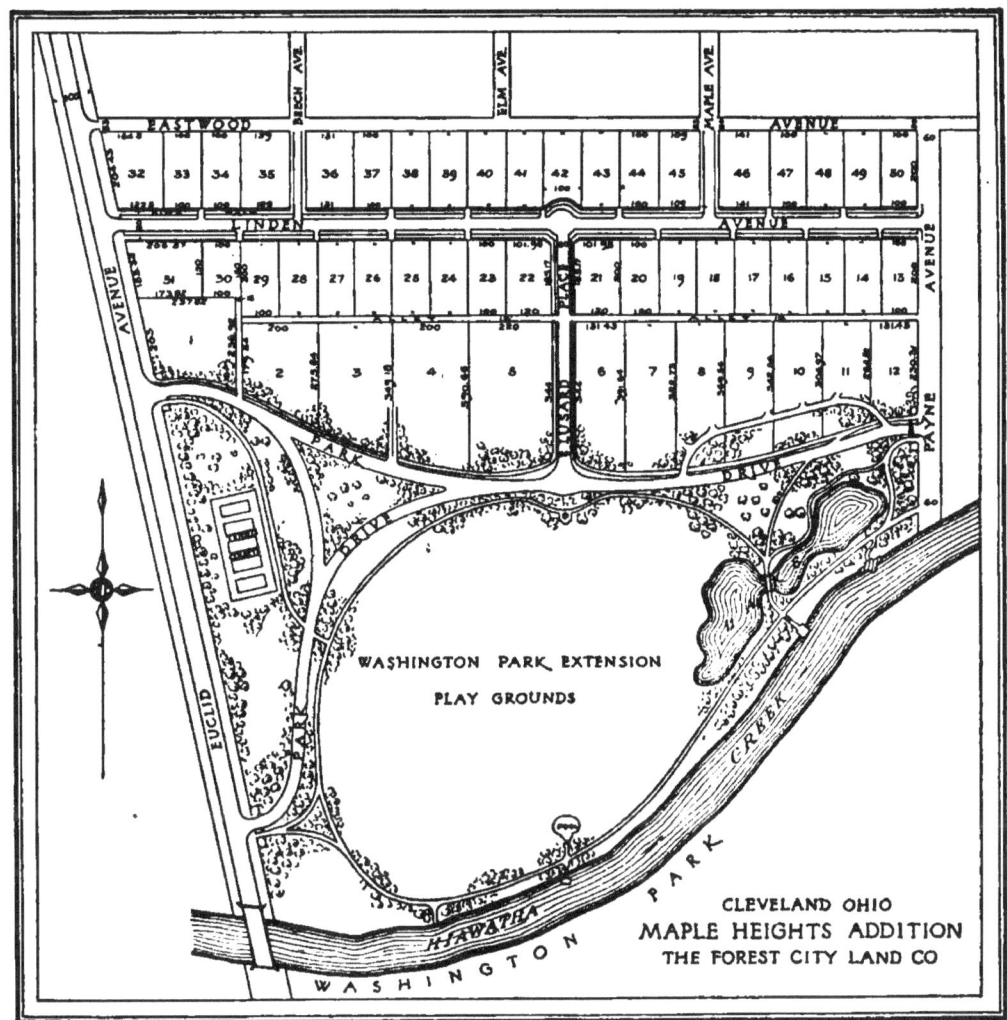

Fig. 145.—Landscape map.

35	Survey of Truck Farm of A.G.Harter Oswego, Ohio.			P.T.Bailey, Surveyor J.H.Pugh, Chain	Tape 24 Compass 17	36
Sta.	Bearing	Distance		July 26.1915	Fair	
A	N72°18'E	760.32	Iron Pin in Concrete	Monument cor. A.Harter, J.Brown, A.Lloyd.		
B	N72°28'E	1149.72	Iron pipe driven in ground.			
C	S7°26'E	1894.20	Stone monument	℄ of Highway		
D	S79°28'W	1074.48	Iron peg ℄ of Highway			
E	N34°00'W	1805.76	Large oak tree blazed. Spike driven in blaze.			
			All bearings taken to right at Sta. P.T.B.			

Fig. 146.—Page from field book.

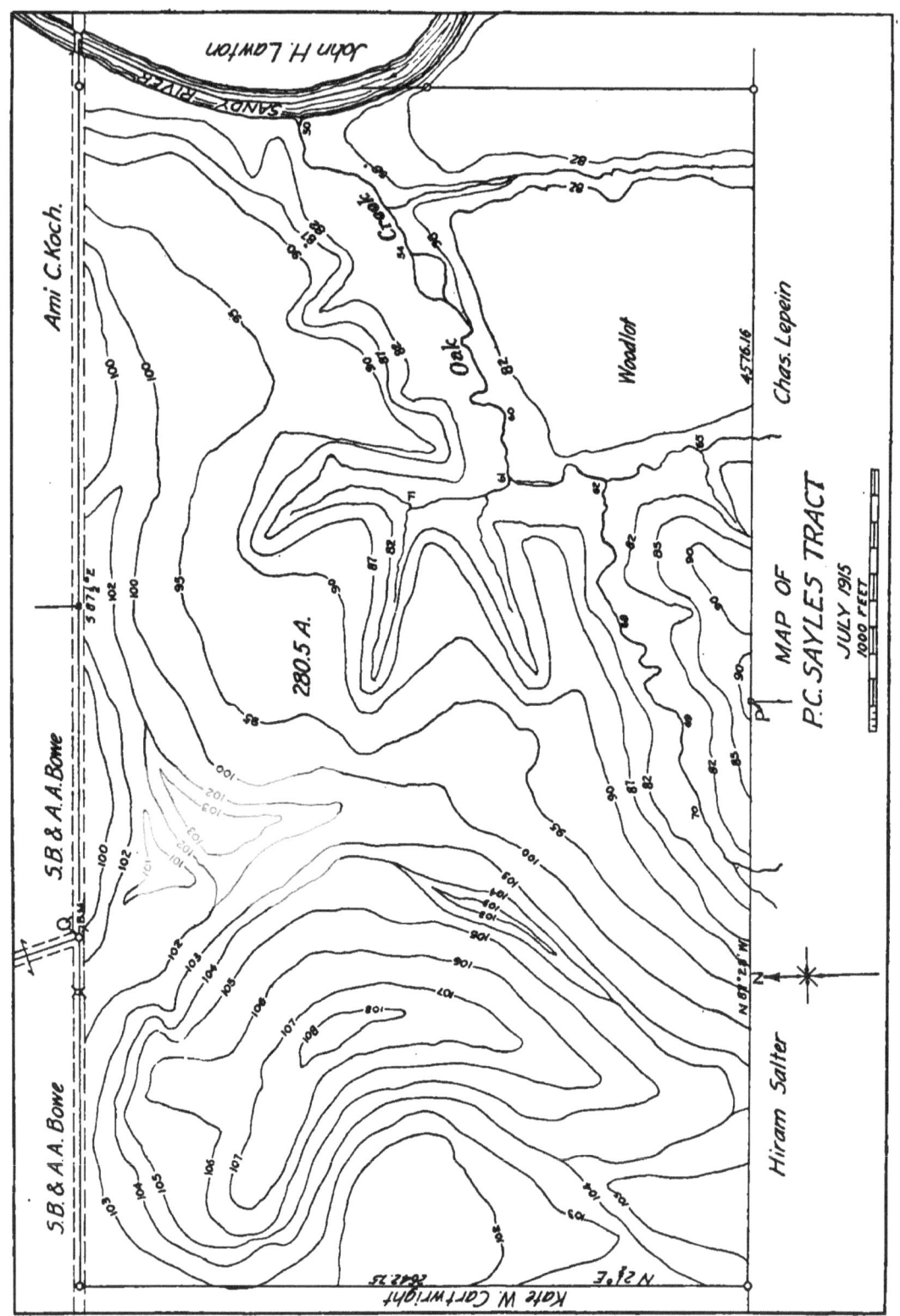

Fig. 147.—Contour map of farm.

PROBLEMS.

1. From the note book page shown in Fig. 146, plat the farm, recording all data as outlined on page 91.

2. Draw a map of your home farm, from deeds, or measurements, showing location of all buildings, and natural and cultural features.

3. Redesign the above, to secure if possible, better and more economical arrangement of plant.

4. From the contour map shown in Fig. 5, page 4, locate the best place for a drain or drains. Draw the map, locating each stretch, and indicate the grade of each. To obtain the grade, draw on Plate A profile paper, a profile of the chosen location. On profile show gradient and depth of ditch at each 50 feet of length. Locate these 50 ft. points or "hubs" on the location map.

5. Locate on the contour map, Fig. 147, a farmstead, bearing in mind drainage, proximity to highway, accessibility to fields, and a prevailing westerly wind.

6. It is proposed to build a 40 ft. road from P to Q, Fig. 147. Choose a location bearing in mind grades, natural obstacles, divisions of ownership etc., and draw a strip map 300 feet wide showing the proposed road. Draw a profile, showing all drains, bridges, culverts, etc.

7. Draw a farm map for office use, of Fig. 147, showing all the items referred to under this head. The western half of this farm needs drainage.

8. On Fig. 147 locate a dam at an advantageous position. Plot the impounding reservoir.

9. Enlarge a portion of Fig. 141, and make a finished drawing, showing planting of different fields by symbols.

10. Enlarge the farmstead of Fig. 5 to the scale of 20 ft. = 1 inch, and draw a landscape map, with proposed planting, walks and drives.

CHAPTER VI

PICTORIAL DRAWING

In Chapter I a general division was made between pictorial drawings and working drawings. Perspective drawing was defined, and reference was made to the simpler pictorial methods of isometric and oblique drawing, which are designed to combine the advantages of both orthographic and perspective. Although they have disadvantages and limitations, familiarity with these two methods is very desirable, as they are often used both to illustrate some object or detail more clearly and to make working sketches and drawings. Working drawings, as we have seen, are generally made in orthographic projection, but often a working drawing of simple construction may be drawn to better advantage in isometric or oblique.

Although theoretically these two systems are somewhat different, the method of construction is very much the same for both.

In orthographic projection we had virtually a separate view for each face of a rectangular object. In these pictorial representations the object is so placed that three faces of it are visible on one view. To avoid confusion each system will be explained separately, and the student should be careful not to confuse the two methods in the same drawing.

Isometric Drawing.

Isometric drawing is based on a skeleton of three lines at 120 degrees apart, called the isometric axes. One is drawn vertically, the others with the 30° triangle, as shown in Fig. 148(*A*). When it is desired to show the

under side of an object the axes are reversed, as at (*B*). The intersection of these three lines would be the front corner of a rectangular object. If the length, breadth and thick-

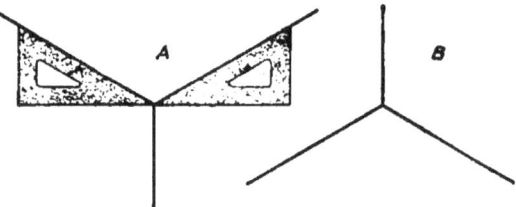

Fig. 148.—Isometric axes.

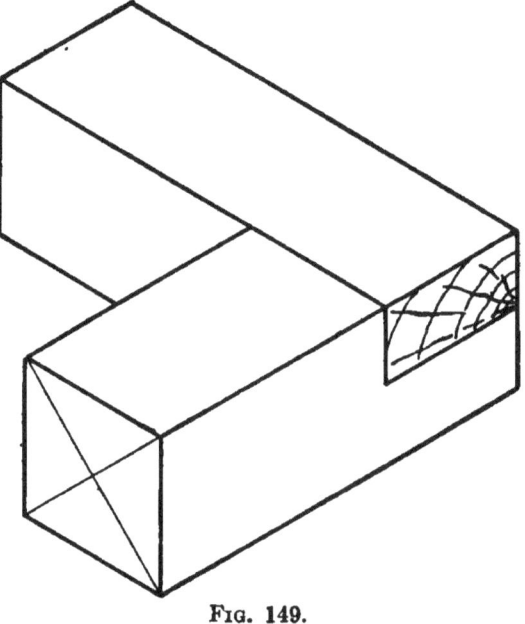

Fig. 149.

ness of the object be laid off on the three axes, the figure may be completed by drawing through these points, lines parallel to the axes, Fig. 149.

A line parallel to an isometric axis is

100

called an isometric line. A line which is not parallel to an isometric axis is called a non-isometric line. The one important rule is **measurements can be made only on isometric lines.** Thus all the lines of an object which has square corners can be object composed entirely of isometric lines, and illustrates the method of making measurements on the original isometric axes and on lines parallel to them.

To draw an object which has non-isometric lines in it, a skeleton of isometric construc-

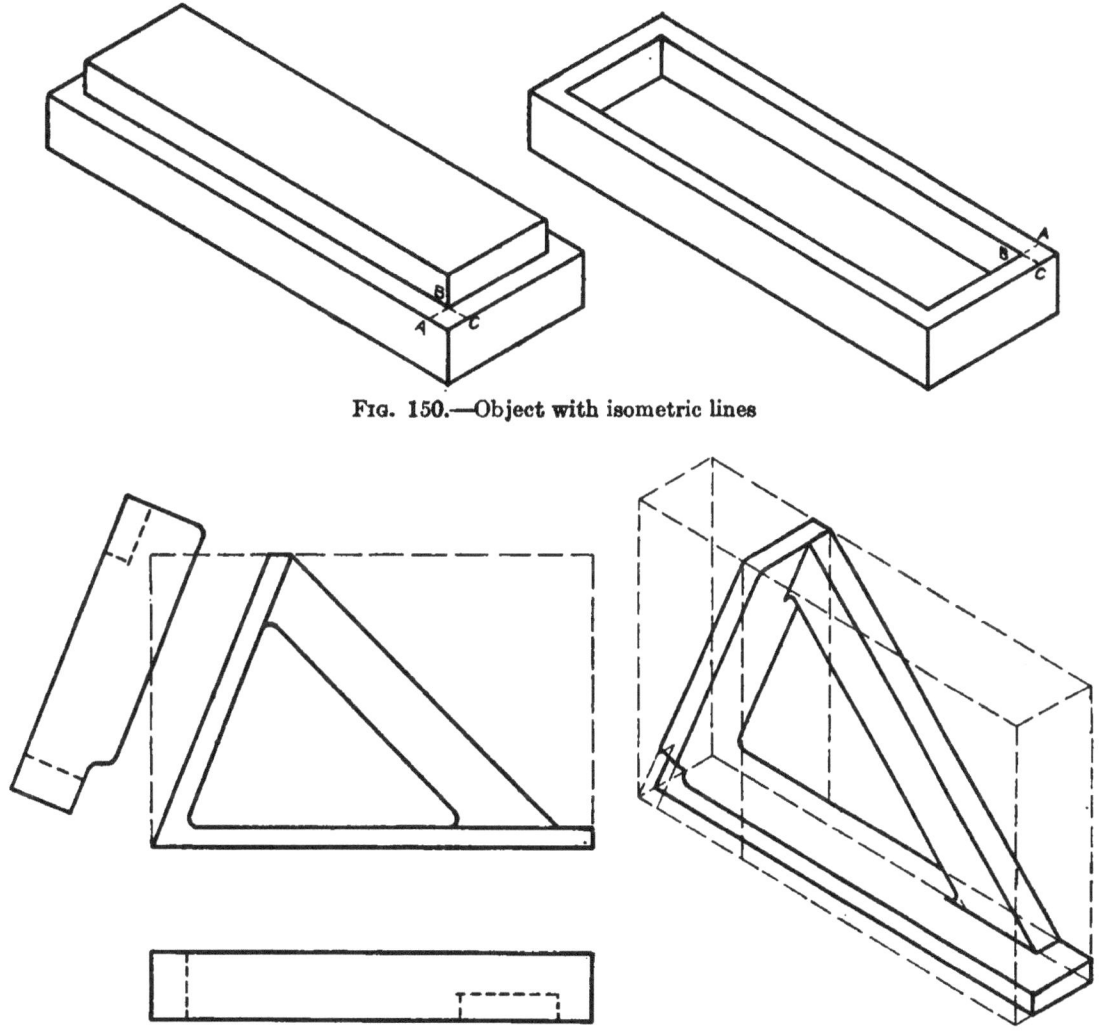

Fig. 150.—Object with isometric lines

Fig. 151.—Object with non-isometric lines.

measured directly. Non-isometric lines can not be measured. For example, the diagonals of the rectangle on Fig. 149 are of equal length on the object but evidently of very unequal length on the isometric drawing.

Fig. 150 is the isometric drawing of an tion lines must be built up, upon which measurements can be made. Often it is necessary to draw one or more orthographic views first and box them with rectangular construction lines, which can be drawn isometrically and measurements upon them made and transferred. Fig. 151 illustrates this construction.

To draw intelligently in isometric it is only necessary to remember the direction of the three principal isometric planes, represented by the three visible faces of a cube or

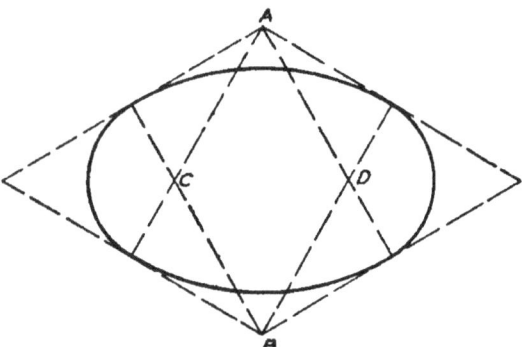

Fig. 152.—Approximate isometric circle.

rectangular figure. Hidden lines are always omitted except when necessary for the description of the piece.

A circle on any isometric plane will be projected as an ellipse, and the isometric square circumscribing the circle must always

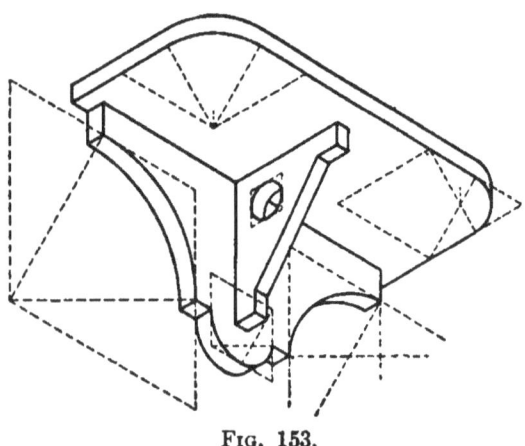

Fig. 153.

be drawn first. The usual construction is to make a four centered circle-arc approximation, finding the centers for the arcs by drawing 60 degree lines from the corners A and B to the middle of the opposite sides, giving centers A, B, C, and D for the ellipse, as shown in Fig. 152. Thus to get the isometric

of any circle-arc the isometric square of its diameter should be drawn in the plane of the face with as much of this construction as is necessary to find centers for the part of the circle needed. Fig. 153 illustrates this construction on the isometric of a shelf, which has been drawn with reversed axes to show the under side.

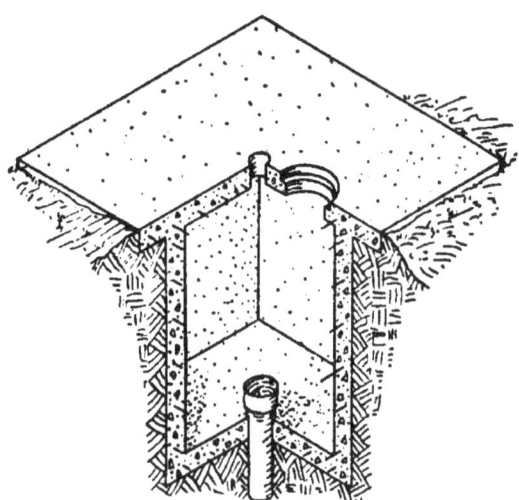

Fig. 154.—An isometric section.

Isometric sections as illustrated in Fig. 154 may sometimes be used to good advantage, the cut faces always being taken in isometric planes.

Oblique Drawing.

Oblique drawing is similar to isometric in having three axes representing three mutually perpendicular lines upon which measurements can be made. It differs in that two of the axes are always at right angles to each other, while the third, or cross axis may be at any angle, Fig. 155, though 30 degrees is generally used. Thus, the face drawn on the plane of the two axes at right angles will appear without distortion, a circle on it for example showing its true shape.

This gives oblique drawing a distinct advantage over isometric for the representation of objects with curved or irregular outlines, and the first rule would be, **place** the object with the irregular outline or contour on the front plane. Fig. 156.

If there is no irregular outline the second

the apparent distortion, which is noticeable both in oblique and isometric drawings, objects in oblique drawing always appearing

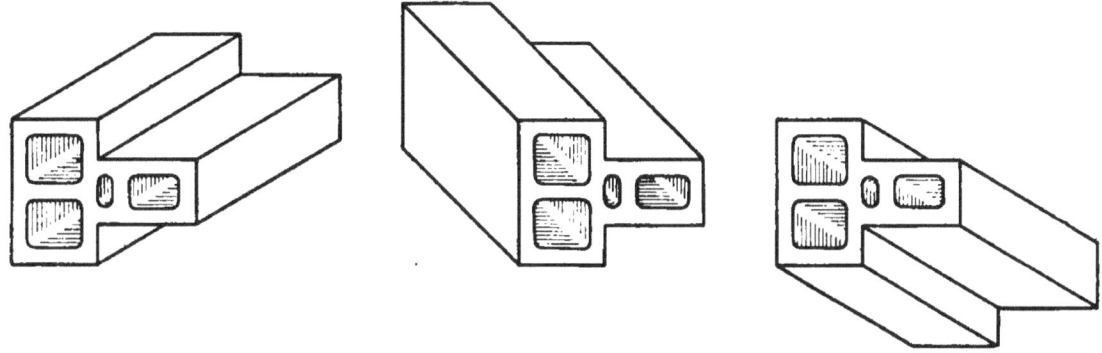

FIG. 155.—Oblique axes illustrated.

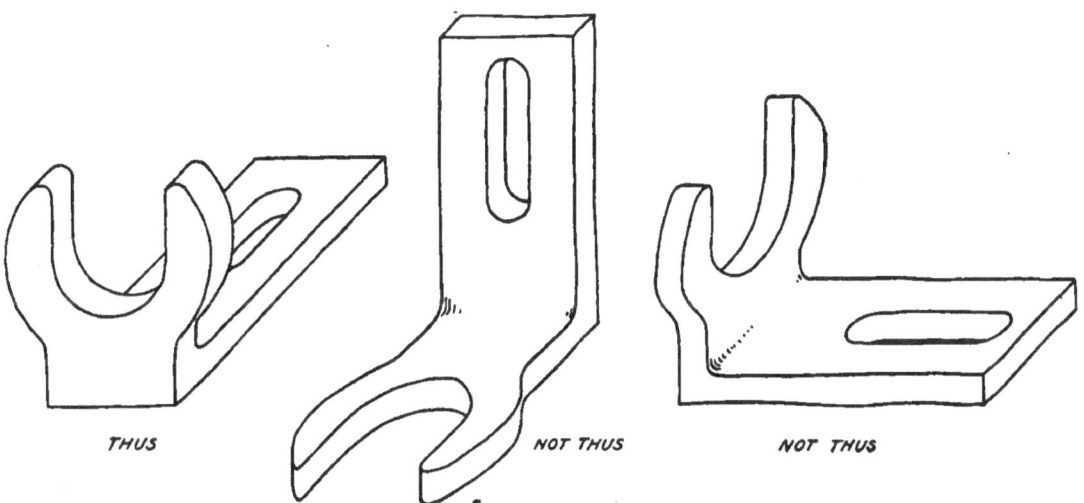

THUS   *NOT THUS*   *NOT THUS*

FIG. 156.—Illustration of first rule.

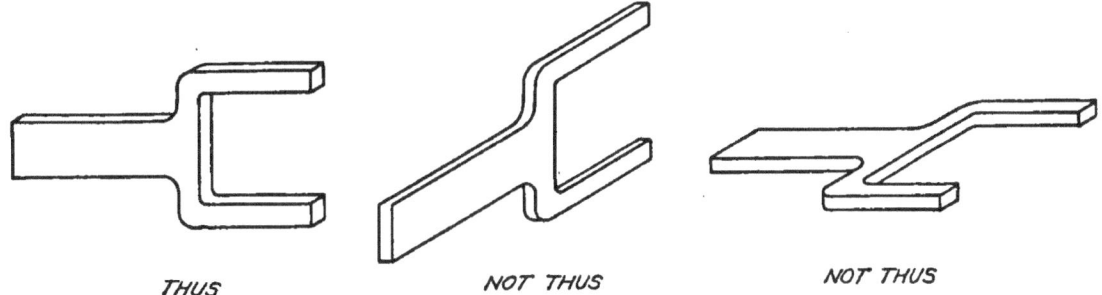

THUS   *NOT THUS*   *NOT THUS*

FIG. 157.—Illustration of second rule.

rule should be followed—**always place the object with the longest dimension parallel to the front,** Fig. 157. This is on account of

to be too wide or thick. Other rules for drawing in oblique are similar to those for isometric.

Cabinet Drawing.

Cabinet drawing is a modification of oblique drawing in which all of the measurements parallel to the cross axis are reduced one-half, in an attempt to overcome the appearance of excessive thickness. Fig. 158 illustrates the comparative appearance of a figure drawn in isometric, oblique and cabinet drawing.

The numerous examples of pictorial drawing used in illustration and explanation throughout this book will serve to suggest various practical applications of the subject that may be made after the student has acquired a facility in its use.

Keep dimension and extension lines in the plane of the face.

Do not confuse the drawing with dotted lines.

Problems.

The following problems are intended to serve two purposes:

1st, for practice in isometric and oblique drawing.

2nd, for practice in reading orthographic projection.

In reading a drawing remember that a line on any view always means a corner or edge and that one must always look at the other

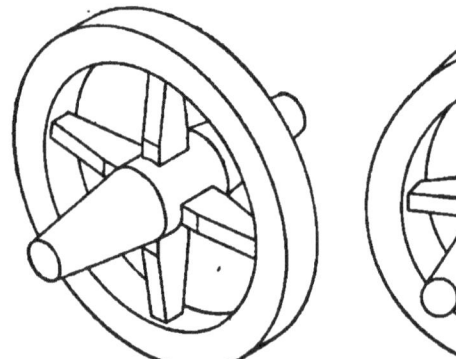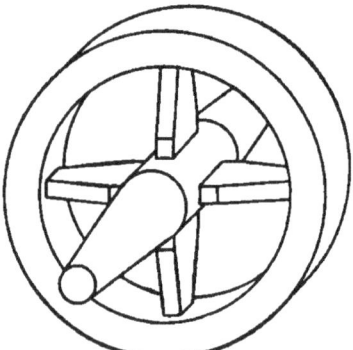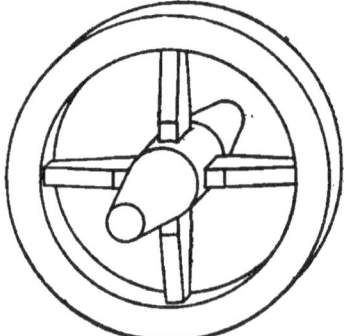

FIG. 158.—Isometric, oblique, and cabinet drawing compared.

Sketching.

One of the valuable uses of isometric and oblique is in making freehand sketches, either dimensioned to form working sketches, or for illustrating some object or detail of construction. The following points should be observed.

Keep the axes flat. The beginner's mistake is in spoiling the appearance of his sketch by getting the axes too steep.

Keep parallel lines *parallel*.

Always block in squares before sketching circles.

In isometric drawing remember that a circle on the top face will be an ellipse with its axis horizontal.

view to find out what kind of a corner it is. Do not try, nor expect to be able, to read a whole drawing at one glance.

The sketches are dimensioned, and it is expected that some will be made as finished drawings, with instruments, while others may be taken as *reading* problems to be sketched freehand. This translation from orthographic to pictorial form is, it will be noticed, just the reverse of what was done in our preliminary study of working drawings, when, in Figs. 19, 20 and 21 we made orthographic projections from the pictures.

To the dimensions and scales given, a 12″ × 18″ sheet will contain any four of the first sixteen problems, thus a sheet may be

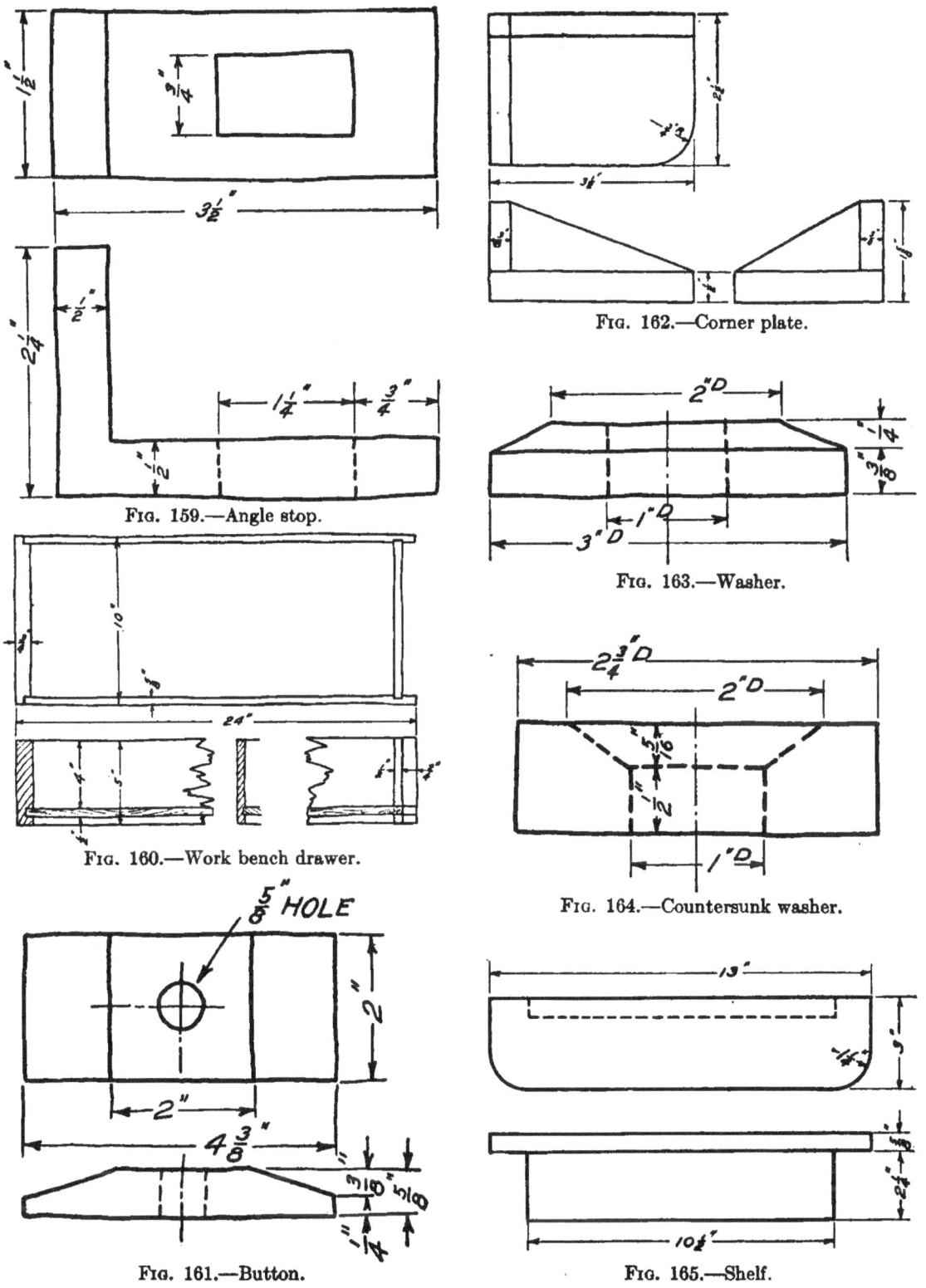

FIG. 159.—Angle stop.

FIG. 160.—Work bench drawer.

FIG. 161.—Button.

FIG. 162.—Corner plate.

FIG. 163.—Washer.

FIG. 164.—Countersunk washer.

FIG. 165.—Shelf.

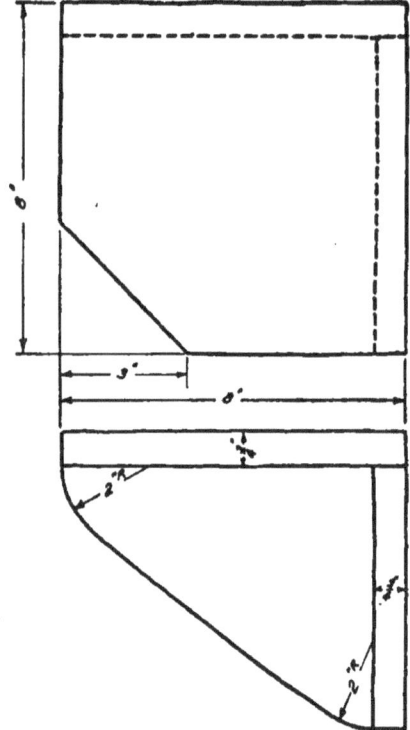

FIG. 166.—Corner bracket.

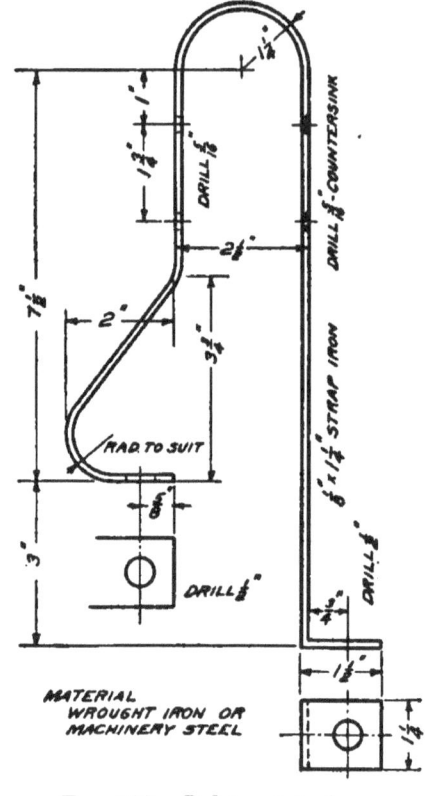

MATERIAL
WROUGHT IRON OR
MACHINERY STEEL

FIG. 168.—Bolster stake iron.

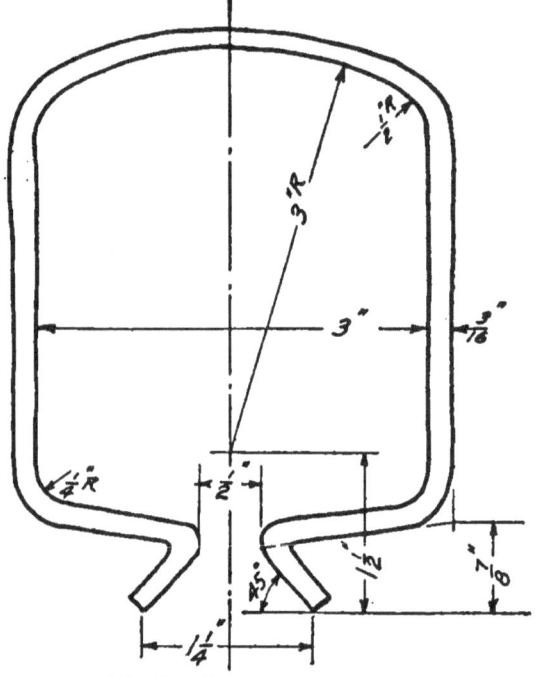

LENGTH TO SUIT SPACE

FIG. 167.—Section of barn door track.

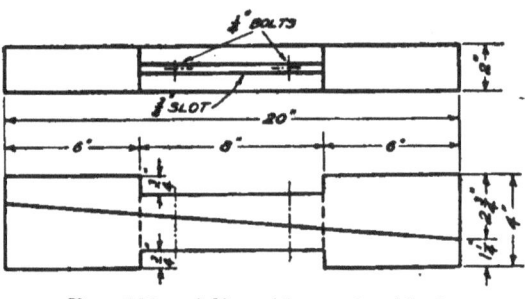

FIG. 169.—Adjustable spacing block.

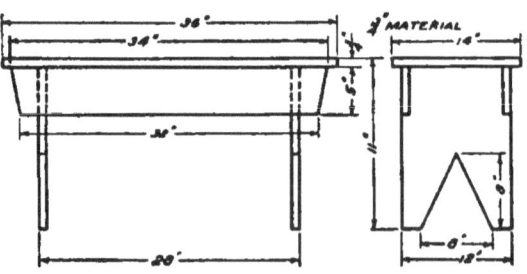

FIG. 170.—Bench.

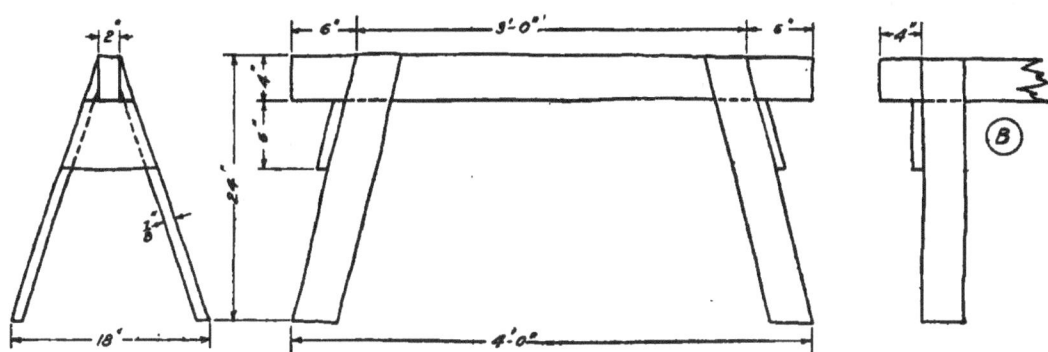

FIG. 171.—Two forms of saw horse.

made up of Nos. 1, 3, 5 and 7, or an alternate of 2, 4, 6 and 8. Selections from problems

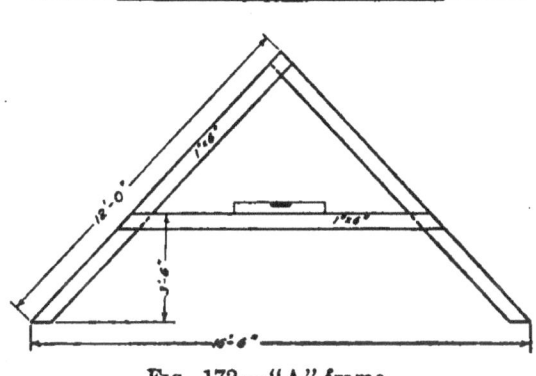

FIG. 172.—"A" frame.

17 to 22 may be placed two on a sheet, and 23 should preferably be made large enough to occupy a sheet alone. Numerous other

pictorial problems such as 24 and 25, may be made from the problems at the end of Chapter III, selecting scale to fit the space allowed.

PROBLEMS.

1. Isometric drawing of angle stop, Fig. 159. Full size.

2. Isometric drawing of drawer, Fig. 160. Scale $1\frac{1}{2}'' = 1'$.

3. Isometric drawing of button, Fig. 161. Full size.

4. Isometric drawing of corner plate, Fig. 162. Full size.

5. Isometric section of washer, Fig. 163. Full size.

6. Isometric section of washer, Fig. 164. Full size.

7. Isometric drawing of shelf, Fig. 165. Scale $6'' = 1'$. Use reversed axes so as to show under side.

8. Isometric drawing of corner bracket, Fig. 166. Scale $3'' = 1'$. Use reversed axes.

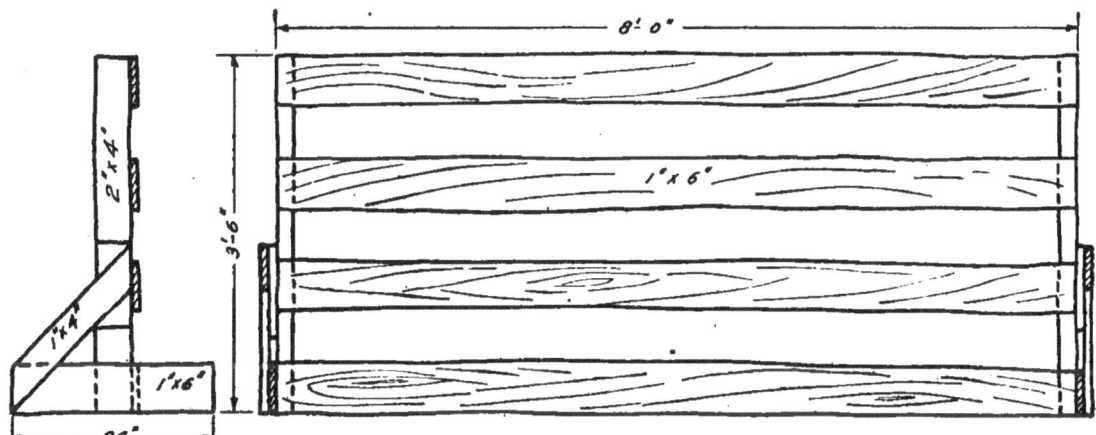

FIG. 173.—Sheep hurdle.

9. Oblique drawing of section of bird-proof track for rolling door. Fig. 167. Full size. Have break line follow form of section.

10. Oblique drawing of bolster stake iron, Fig. 168. Scale 6″ = 1′.

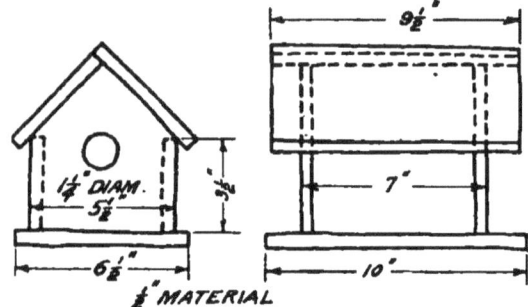

FIG. 174.—Bird house.

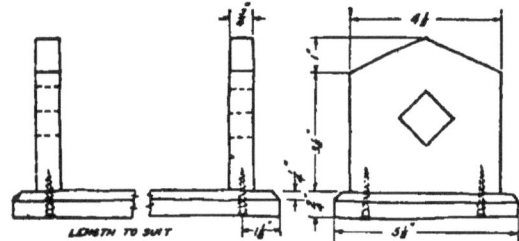

FIG. 175.—Book rack.

11. Oblique (or isometric) drawing of spreader for concrete forms. Fig. 169. Scale 3″ = 1′.

12. Oblique (or isometric) drawing of bench, Fig. 170. Scale 1½″ = 1′.

13. Isometric drawing of horse, Fig. 171. Scale 1″ = 1′. Notice construction for non-isometric lines.

14. Isometric drawing of horse, alternate form B of Fig. 171. Scale 1″ = 1′.

15. Oblique drawing of land level frame, Fig. 172. Scale ⅜″ = 1′.

16. Oblique (or isometric) drawing of sheep hurdle, Fig. 173. Scale ½″ = 1′.

17. Oblique (or isometric) drawing of bird house, Fig. 174. Scale 6″ = 1′.

18. Oblique (or isometric) drawing of book rack. Fig. 175. Scale 6″ = 1′.

19. Oblique (or cabinet) drawing of wheel, Fig. 176. Scale 3″ = 1′.

20. Isometric (or oblique) drawing of angle brace, Fig. 177. Full size.

21. Isometric (oblique or cabinet) drawing of kitchen step ladder, Fig. 178. Scale 3″ = 1′.

22. Isometric (or oblique) drawing of stone rake, Fig. 179. Scale ¾″ = 1′.

23. Oblique (isometric or cabinet) drawing of two-till tool chest, Fig. 180. Scale 3″ = 1′. Draw with lid open, or break out a section, in order to show interior.

24. Isometric or oblique drawing of milking stool, Fig. 75.

25. Isometric or oblique drawing of gang mold, Fig. 77.

The objects of Fig. 181 are to be sketched freehand in the most suitable pictorial system.

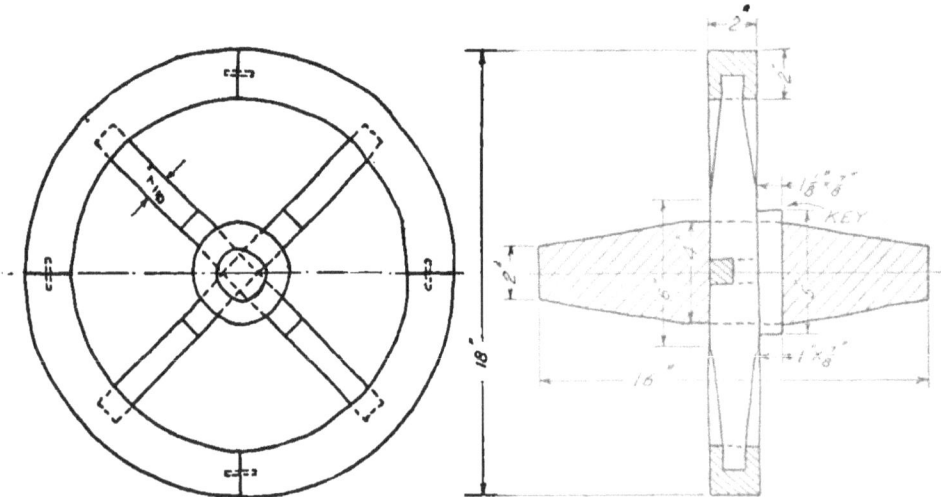

FIG. 176.—Wooden wheelbarrow wheel.

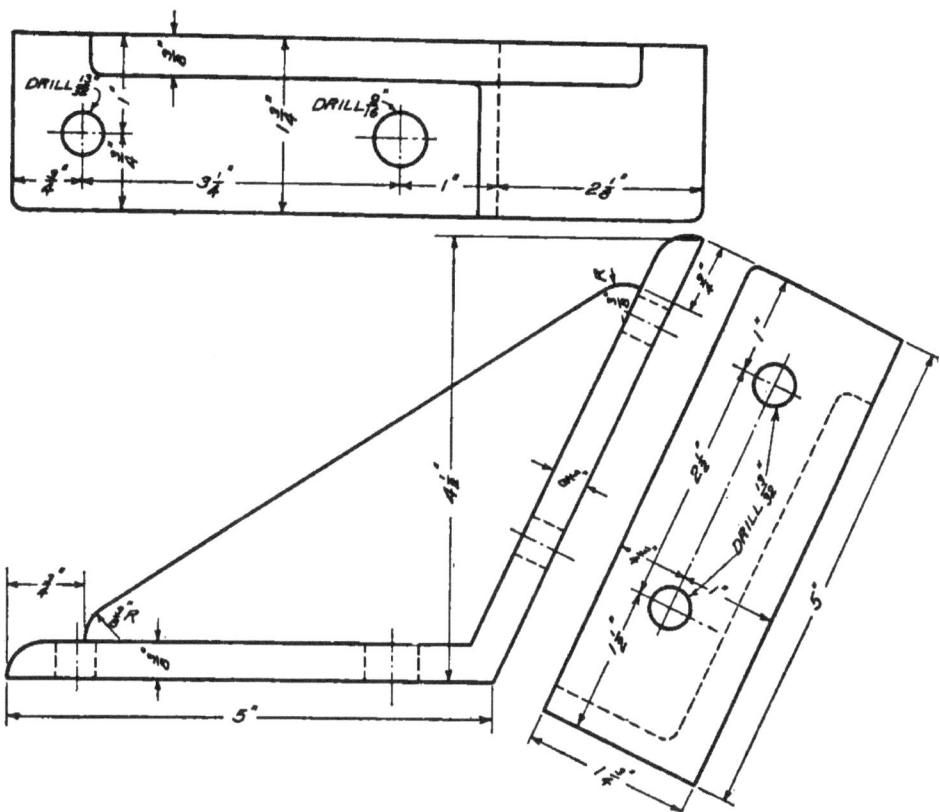

Fig. 177.—Angle brace.

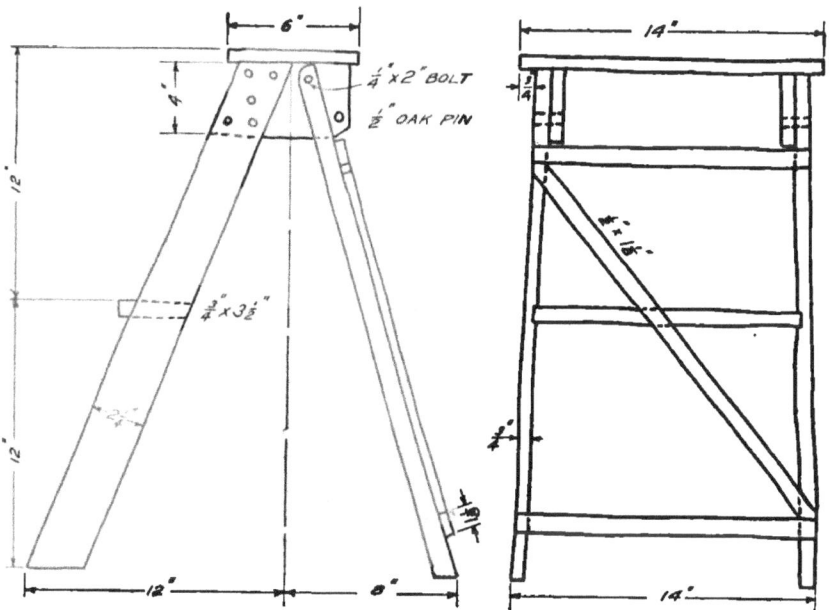

Fig. 178.—Kitchen step-ladder.

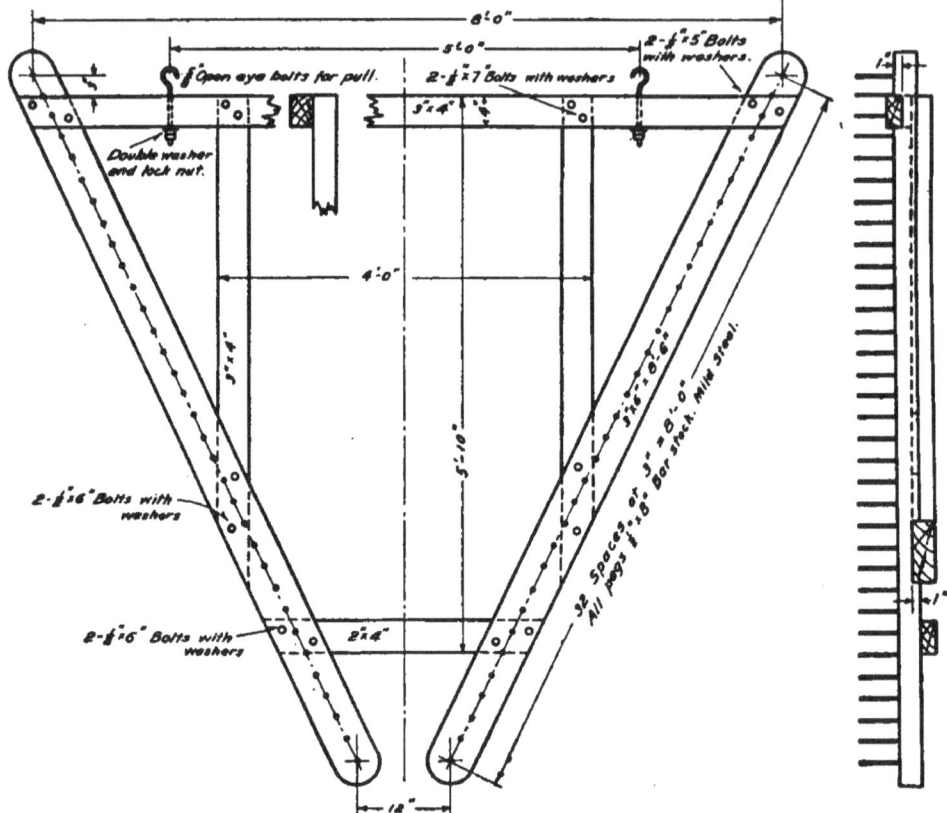

Fig. 179.—Stone rake.

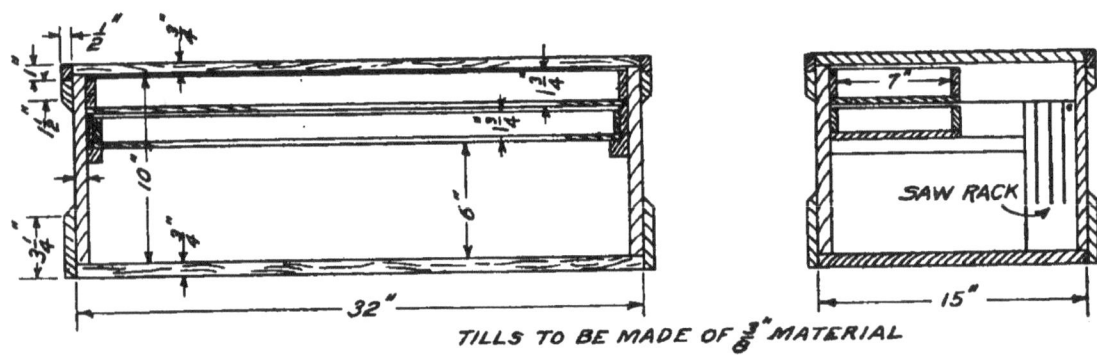

TILLS TO BE MADE OF ⅜" MATERIAL

Fig. 180.—Tool chest.

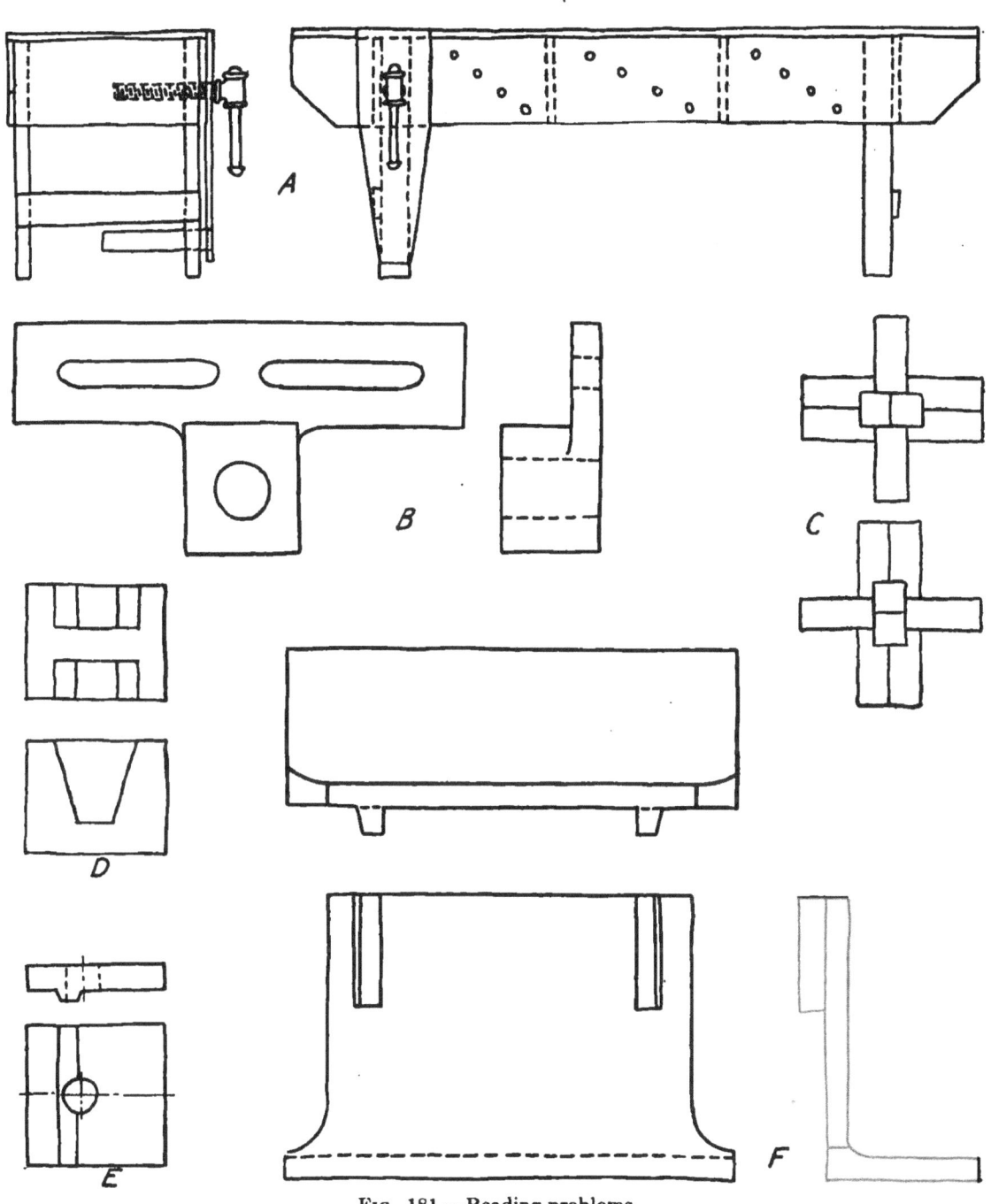

Fig. 181.—Reading problems.

CHAPTER VII

CONSTRUCTION DATA

In designing structures, and preparing working drawings and specifications, the designer needs certain information as to materials and commercial practice. A number of tables, and items of miscellaneous information are assembled in this chapter, for ready reference; and as suggestions in their use a few problems are appended.

Stock and Commercial Sizes.

For economy, and avoidance of trouble in procuring, stock sizes of materials should be used as far as possible. As these commercial sizes and standards vary somewhat in different parts of the country, one should be sure that grades and sizes of any materials specified are locally recognized, and obtainable.

Lumber.

Lumber comes in standard lengths of 10–0″, 12′–0″, 14′–0″, etc., up to 24′–0″. Over 24′–0″ and under 10′–0″ is special.

In width and thickness the following nominal sizes are obtainable.

Thickness					Width	
1″..............	4″,	6″,	8″,	10″,	12″	
2″..............	4″,	6″,	8″,	10″,	12″	
3″ (local)........	4″,	6″,	8″,	10″,	etc., to 16″	
4″..............	4″,	6″,	8″,	10″,	12″,	14″
6″..............	6″,	8″,	10″,	12″,	14″,	16″
8″..............	10″,	12″,	14″			
10″..............	10″,	12″,	14″,	16″		
12″..............	12″,	14″,	16″			
14″..............	14″,	16″				

These sizes are called nominal, as the actual measurement varies with different woods and in different localities. A piece of 2″ × 4″ Y.P., for example, may be not larger than $1\frac{5}{8}$″ or $1\frac{3}{4}$″ × $3\frac{3}{4}$″.

Dressed lumber is indicated S1S, S2S, S4S, according as it is surfaced on one side, both sides, or all over. A 1″ S1S board is thus $1\frac{5}{16}$″ thick, and an S2S, $\frac{7}{8}$″ thick.

Shingles are termed $\frac{5}{2}$ or $\frac{6}{2}$ according as 5 butts or 6 butts measure 2″ thick. They are 16″ long, and there are 4 packs to M (the letter M is always used for 1000.).

Mill Work means work prepared at the planing mill. It is charged for by the linear foot, or by the piece.

Doors have standard thicknesses of $1\frac{1}{8}$″, $1\frac{3}{8}$″ and $1\frac{3}{4}$″. Their widths and heights vary in even two inches.

Sash has the same standard thicknesses as doors, and is specified by the glass sizes. To obtain the width of frame add $4\frac{1}{8}$″ to the glass width for a 2-light window, and for the length add 6″ to the length of the two panes.

Windows are made either "check rail," "plain rail," or "casement." An order would read 2 Lt, 24″ × 24″ × $1\frac{3}{4}$ D.S.

Glass comes in single and double strength window glass, plate glass of different thicknesses, and a variety of obscure glass, such as "frosted," "chipped," "Florentine," etc. Sizes run in inches as 6 × 8; 7 × 9; 8 × 10; 8 × 12; etc.

Sheet Metal is sold by gauge number 10 to 30 (thick to thin). Galvanized iron runs in size from 24″ × 72″, the widths advancing by 2″ to 30″, then by 6″ to 48″, and the lengths by 12″ to 120″. Tin comes in two weights IC and IX, of which the latter is heavier, and in sizes 20″ × 28″ and 14″ × 20″.

Wire is measured by wire gauge number, 00 to 36, thick to thin.

Pipe is measured by the nominal inside diameter, as explained on page 32. It comes in lengths of about 20 ft. and either plain or galvanized. Screws, nails, etc., are too various to be enumerated. One not familiar with the trade sizes and classifications should at once become so, with the assistance of a hardware store man.

Rope is measured by its largest diameter, and specified by its number of strands, as 3 strand or 4 strand. The best sash cord is braided.

Drain Tile comes in lengths 12″ and 24″. The smallest practical size is 3″, and diameters advance by inches up to 10″ then by 2″ up to 24″.

Slate is sold by the square, 100 sq. ft. There are various kinds and grades. Widths advance by inches and lengths by two inches.

Metal Roofing comes in sheets or rolls, the sizes the same as galvanized iron.

Ready Roofing comes under various trade names, and is usually a composition of bituminous or resinous compounds and felt or paper. It is sometimes coated with coarse sand or slate. It is sold in rolls of 108 sq. ft., making one square with allowance for laps.

Weight of Roofing.

The approximate weights of different roof coverings are needed in determining the strength of rafters. The snow load, and wind pressure are also factors to be added in figuring roofs, and an important problem should not be attempted without assistance.

Shingles run about 250 lbs. per square. Slate 750 to 900, tile 1100 to 1400, ready roofing 40 to 150, galvanized iron 100 to 300.

Weights of Materials.

The weight per cubic foot of different materials is often required. The following table gives the weight of some of the common materials.

	Lbs. per cu. ft.
Oak	52
Yellow pine	45
White pine	25
Hemlock	25
Maple	49
Brick (average)	125
Concrete (average)	140

	Lbs. per cu. ft.
Sand, dry	95
Sand, wet	130
Gravel	130
Cast iron	450
Wrought iron	480
Granite	170
Ice	53
Water	62.5

A gallon of water weighs 8⅓ lbs. and contains 231 cu. inches.

Space Required for Storage.

Hay, one ton occupies 512 cu. ft.
Straw, one ton occupies 600 to 800 cu. ft.
Hay, pressed, 25 lbs. per cu. ft.
Small grain, 1 bu. = 1¼ cu. ft.
Corn on cob, 1 bu. = 2½ cu. ft.

Space Required for Farm Implements.

The following table gives the over-all allowance for some common implements. Add to the list for your own requirements.

Side delivery rake	10½′ × 12′	
Corn planter	6′ × 5′	
Mower	6′ × 6′	
Hay loader	9′ × 10½′ × 10′	
Rake (common 2 horse)	5′ × 11′	
30 H.P. tractor	8′ × 17′ × 12′	
Corn cultivator	5′ × 7′	
Corn binder	9′ × 14′	
Grain binder (7 ft. cut)	9′ × 14′	
Manure spreader	8′ × 16′	
Grain drill (8 hoe)	5′ × 9′	
Roller	8′ × 3′	

Ration for Beef Feeders.

12 lbs. shelled corn
2 lbs. cotton seed meal
20 lbs. corn silage
5 lbs. clover or alfalfa
per 1000 lbs. weight of live animal.

Table for the Selection of Native Woods.

The following list gives the suitable woods for different purposes, arranged somewhat in the order of their desirability and availability.

Light framing—hemlock, poplar, spruce, oak.
Heavy framing—yellow pine, oak.
Joists—yellow pine, oak, redwood.
Siding—cypress, redwood, yellow pine, poplar.
Shingles—cypress, red cedar, redwood, white cedar, pine.

Flooring—oak, maple, edge grain yellow pine, birch (cypress for porch).

Interior finish—oak, birch, cypress, yellow pine; cherry, walnut, mahogany, etc.

Tanks—cypress, redwood, fir.

Fence posts—(in order of durability) osage orange, black locust, red cedar, mulberry, white cedar, catalpa, chestnut, oak, black ash.

Vehicles—hickory, oak, ash.

Strength of Timbers.

The following tables give the safe load in pounds that can be carried by long leaf yellow pine timbers, as figured from the formulas indicated. To find the comparative strength of other kinds of wood take 75% of the amount for white oak; 60% for white pine and 55% for hemlock.

TABLE I

Beam fixed at one end and concentrated load at the other end. Formula $W = \dfrac{BD^2A}{4L}$, where W is the safe load in lbs., B the breadth of the beam in inches, D its depth in inches and L its length in feet. A is the constant, 100 for yellow pine, and as indicated above, 75 for white oak, etc.

B, in.	D, in.	L, feet	W, pounds
2	4	4	200
2	6	4	450
2	8	4	800
3	4	4	300
3	6	4	675
3	8	4	1200
3	8	6	800
4	4	4	400
4	6	4	900
4	6	6	600
4	6	8	450
4	8	6	1066
4	8	8	800
6	6	6	900
6	6	8	675
6	6	10	540
6	8	6	1600
6	8	8	1200
6	8	10	960
6	10	10	1500
8	10	10	2000
8	10	12	1666
8	12	10	2880
8	12	12	2400
10	12	10	3600

TABLE II

Beam fixed at one end and uniformly loaded. Formula $W = \dfrac{BD^2A}{2L}$

B	D	L	W
2	4	4	400
2	8	4	1600
*(Continued.)			

*The values of W in this table are evidently just double those of Table I.

TABLE III

Beam supported at both ends and uniformly loaded. Formula $W = \dfrac{2BD^2A}{L}$

B, in.	D, in.	L, feet	W, lbs.
2	4	8	800
2	4	10	640
2	4	12	532
2	6	12	1200
2	8	12	2132
2	10	12	3332
2	10	14	2856
*2	10	16	2500
*10	2	4	2000
2	12	12	4800
3	8	12	3200
3	10	12	5000
3	12	12	7200
3	12	16	4800
4	8	12	4266
4	10	12	6666
4	12	12	9600
6	8	12	6400
6	10	12	10000
6	12	12	14400
8	8	12	8532
8	10	12	13332
8	12	12	19200
8	12	16	14400
10	12	12	24000
10	12	16	18000

* Note difference between edgewise and flat.

TABLE IV

Beam supported at both ends, with concentrated load in the middle. Formula $W = \dfrac{BD^2A}{L}$

B	D	L	W
2	4	10	320
2	4	12	266
* (Continued.)			

* The values of W in this table are evidently one-half of those in Table III.

Concrete.

The following table will serve as a guide as to the proportions of cement, sand and stone used in making concrete for different purposes. It also gives the amounts of each in a cubic yard of concrete of the different mixtures, which is necessary to have in figuring the quantities needed for a piece of work.

TABLE V

Proportions			Cement, bbl.	Sand, cu. yd.	Stone, cu. yd.	Use
Cement	Sand	Stone				
1	2	3	1.70	0.52	0.77	(A)
1	2	4	1.46	0.44	0.89	(B)
1	2½	5	1.19	0.46	0.91	(C)
1	3	6	1.01	0.46	0.92	(D)
1	4	8	0.77	0.47	0.93	(E)

(A) 1:2:3 for tanks; silos; reinforced beams and columns; fence posts.

(B) 1:2:4 for concrete blocks; piers; silos; small culverts; foundations; single course walks.

(C) 1:2½:5 for stable floors; foundation walls (light).

(D) 1:3:6 for footings; barnyard paving; first course of two course walks.

(E) 1:4:8 for footings; mass work where great strength is not required.

(F) Top course of two-course floors and walks made of 1:2 cement and sand.

Cement is usually sold in sacks of 96 lbs. or 4 sacks = 1 bbl.

Silos and Silage.

The following table gives the capacities in tons of silos of different sizes; the approximate number of acres required to fill each size; the number of cows, on a 40 lb. ration, to keep the ensilage fed down the minimum amount per day; and the capacity in number of cows, based on 180 days feed with 40 lb. ration.

All data refers to settled ensilage.

TABLE VI

Diam., feet.	Height, feet.	Capy., tons	No. acres to fill	No. cows, to feed off 2"	No. cows for 180 days feed
10	20	26	3– 5	13	7
10	30	47	5– 8	13	13
10	32	51	6– 8	13	14
11	32	62	7–10	15	16
12	32	74	8–10	19	21
12	36	87	9–12	19	24
14	30	91	9–12	25	25
14	34	109	11–15	25	30
14	38	128	11–16	25	35
16	32	131	12–17	33	36
16	36	155	15–21	33	43
16	40	180	16–22	33	50
16	44	207	18–26	33	55

Sunshine Table.

Table VII is compiled from Farmers' Bulletin No. 438, U. S. Department of Agriculture and is of use in designing swine houses, in showing the relation between windows and width of pen so that the sun's rays may strike the floor. A is the height of top of window, and B the horizontal distance from window to rear of pen.

Dairy Score Card.

The score card for dairy inspection given on the next page will be of value in checking the plans of a dairy barn, or in examining an existing plant. It is printed in Circular 199, U. S. Department of Agriculture, Bureau of Animal Industry.

DAIRY SCORE CARD

Equipment	Score Perfect	Score Allowed	Methods	Score Perfect	Score Allowed
Cows			**Cows**		
Health..................................	6	....	Clean...................................	8	....
Apparently in good health 1			(Free from visible dirt, 6.)		
If tested with tuberculin within a year			**STABLES**		
and no tuberculosis is found, or if			Cleanliness of stables...................	6	....
tested within six months and all re-			Floor....................... 2		
acting animals removed.......... 5			Walls....................... 1		
(If tested within a year and react-			Ceiling and ledges.................. 1		
ing animals are found and			Mangers and partitions............ 1		
removed, 3.)			Windows..................... 1		
Food (clean and wholesome).............	1	....	Stable air at milking time...............	5	....
Water (clean and fresh).................	1	....	Freedom from dust.............. 3		
STABLES			Freedom from odors.............. 2		
Location of stable.......................	2	....	Cleanliness of bedding...................	1	....
Well drained...................... 1			Barnyard...............................	2	....
Free from contaminating surround-			Clean 1		
ings........................... 1			Well drained................... 1		
Construction of stable...................	4	....	Removal of manure daily to 50 feet		
Tight, sound floor and proper gutter. 2			from stable.........................	2	....
Smooth, tight walls and ceiling..... 1			**MILK ROOM OR MILK HOUSE**		
Proper stall, tie, and manger........ 1			Cleanliness of milk room...............	3	....
Provision for light: Four sq. ft. of glass			**UTENSILS AND MILKING**		
per cow............................	4	....	Care and cleanliness of utensils..........	8	....
(Three sq. ft., 3; 2 sq. ft., 2; 1 sq. ft.,			Thoroughly washed................ 2		
1. Deduct for uneven distribu-			Sterilized in steam for 15 minutes... 3		
tion.)			(Placed over steam jet, or scalded		
Bedding................................	1	....	with boiling water, 2.)		
Ventilation............................	7	....	Protected from contamination...... 3		
Provision for fresh air, controllable			Cleanliness of milking...................	9	....
flue system..................... 3			Clean, dry hands.................. 3		
(Windows hinged at bottom, 1.5;			Udders washed and wiped.......... 6		
sliding windows, 1; other open-			(Udders cleaned with moist cloth,		
ings, 0.5.)			4; cleaned with dry cloth or		
Cubic feet of space per cow 500 ft.....			brush at least 15 minutes before		
(Less than 500 ft., 2; less than 400			milking, 1.)		
ft., 1; less than 300 ft., 0.)					
Provision for controlling tempera-			**HANDLING THE MILK**		
ture........................... 1			Cleanliness of attendants in milk room....	2	....
UTENSILS			Milk removed immediately from stable		
Construction and condition of utensils....	1	....	without pouring from pail.............	2	....
Water for cleaning......................	1	....	Cooled immediately after milking each		
(Clean, convenient, and abundant.)			cow.................................	2	....
Small-top milking pail..................	5	....	Cooled below 50° F....................	5	....
Milk cooler............................	1	....	(51° to 55°, 4; 56° to 60°, 2.)		
Clean milking suits.....................	1	....	Stored below 50° F....................	3	....
MILK ROOM OR MILK HOUSE			(51° to 55°, 2; 56° to 60°, 1.)		
Location: Free from contaminating sur-			Transportation below 50° F.............	2	....
roundings...........................	1	....	(51° to 55°, 1.5; 56° to 60°, 1.)		
Construction of milk room..............	2	....	(If delivered twice a day, allow per-		
Floor, walls, and ceiling............ 1			fect score for storage and trans-		
Light, ventilation, screens.......... 1			portation.)		
Separate rooms for washing utensils and					
handling milk.......................	1	....			
Facilities for steam.....................	1	....			
(Hot water, 0.5.)					
Total..................................	40	....	Total	60	....

Equipment.................+ Methods......................... = ...Final Score.

NOTE 1.—If any exceptionally filthy condition is found, particularly dirty utensils, the total score may be further limited.

NOTE 2.—If the water is exposed to dangerous contamination, or there is evidence of the presence of a dangerous disease in animals or attendants, the score shall be 0.

TABLE VII.—SUN IN REAR AT 10 A. M. AND 2 P. M.

Lat.	Jan. 1st		Feb. 1st		Mar. 1st		Apr. 1st	
	A	B	A	B	A	B	A	B
38°N.	4'-10"	10'-0"	6'-4"	10'-0"	9'-4"	10'	15'-6"	10'
	5'-10"	12'	7'-7"	12'	11'-3"	12'	18'-7"	12'
40°N.	4'-5"	10'	5'-10"	10'	8'-9"	10'	14'-4"	10'
	5'-4"	12'	7'-0"	12'	10'-6"	12'	17'-2"	12'
42°N.	4'-0"	10'	5'-5"	10'	8'-2"	10'	13'-4"	10'
	4'-10"	12'	6'-6"	12'	9'-9"	12'	16'-0"	12'
44°N.	3'-7"	10'	4'-11"	10'	7'-7"	10'	12'-5"	10'
	4'-4"	12'	5'-11"	12'	9'-1"	12'	14'-11"	12'
46°N.	3'-3"	10'	4'-6"	10'	7'-0"	10'	11'-6"	10'
	3'-10"	12'	5'-5"	12'	8'-5"	12'	13'-10"	12'
48°N.	2'-10"	10'	4'-1"	10'	6'-6"	10'	10'-9"	10'
	3'-5"	12'	4'-11"	12'	7'-10"	12'	12'-11"	12'

Kitchen Score Card.

The kitchen score card given may be used in connection with the planning of a new kitchen or in remodeling an old one. In this method of judging, the ideal kitchen would score 100 points, and cuts are made under each of the four heads for any items failing to reach the standard indicated. Some of these depend on actual measurements, while others are matters of judgment.

KITCHEN AT .... SCORED BY	Total	Cut	Score
1. *Plan*—35 Points.			
1—Arrangement of space for equipment...................	15		
Sink—convenience of; Stove—convenience of; Table—convenience of.			
2—Storage....................	15		
Stores Pantry, size, convenience;			

(*Kitchen score card, cont.*).........

	Total	Cut	Score
Serving Pantry, convenience; Refrigerator, convenience; Shelving, adequate; Clock Shelf. Distances—If any two (sink, table, stove, pantry) are farther apart than 12 ft., cut ½ pt. for each ft.			
3—Doors.......................	5		
If more than 4, cut 1 point for each. Outside door direct to covered porch, if no covered porch, cut. Door to Dining Room double swung if direct. Accessibility to front door. Accessibility to upstairs. Accessibility to cellar. If rear stairs go up from Kitchen, 3 pts.			
11. *Light and Ventilation*.............	25		
Two exposures; if only one cut 5 points.			

(*Kitchen score card, cont.*)..........

	Total	Cut	Score
Glass area = 20% of floor area, cut 1 point for each 1% under. Window in Pantry, cut 2 pts. if none. Satisfactory daylight—at stove, at sink, at table, 3 points each. Satisfactory artificial light—at stove, at sink, at table, 3 points each. Transom on outside door, 1 point. Height of sills—If under 34″ cut 1 point. Ventilating hood or flue—1 point.			
III. Floors and Walls—10 Points.			
1—Floor Resilient and grease proof. Hardwood, Monolith or Linoleum, O. K. Cut for cracks, softwood, oil cloth, carpet, etc.	4		
2—Walls......................... Light, cheerful, sanitary. Cut for wall paper or dark color.	4		
3—Woodwork................... Cut for dust catching mouldings or projections—1 point. Cut for wood wainscot—1 point.	2		
IV. Equipment—30 points.			
1—Stove—adequate size and condition...................... If oven is less than 10″ from floor cut 1 pt. per inch. If no broiler, cut 2 points. If no thermometer, cut 1 point.	12		
2—Sink........................ Enamel or Porcelain O. K. Cut for iron, tin, etc., 3 points. Drainboard double, cut 3 points if single. Splashboard, cut 2 points if wood.	8		
3—Table....................... Size—cut 1 pt. if smaller than 6 sq. ft. Height—cut 1 pt. if uncomfortable.	3		
4—Refrigerator................. Size, material, condition, drainage.	4		
5—Fireless Cooker..............	2		
6—Chair and stool.............	1		
Total..................	100		

If no water in Kitchen, cut 40 points.
If no hot water in Kitchen, cut 20 points.
If Kitchen is used as Laundry, cut 15 points.

REMARKS

....
....
....

SUGGESTIONS FOR IMPROVEMENT

....
....
....

Estimating.

In making an estimate of the cost of a proposed structure, there are two general methods, the approximate estimate, and the detailed estimate.

Approximate estimates are used for determining the size possible for a given cost, or for getting the approximate cost after preliminary sketches are made. The methods are based on a knowledge of the cost of similar structures, and are fairly accurate and reliable if all conditions, such as the relative cost of labor and materials, are known.

The **cubic method** is the one generally used. It consists simply in figuring the cubic contents of the entire building, measuring to the outside of the walls, and multiplying by a unit price per cubic foot, which has been determined by dividing the cost of one, or a number of similar structures by the number of cubic feet contained.

The following list of unit costs may be taken as giving safe general averages at the present time. The cost of building is steadily increasing, with the rise in prices of materials and labor, and a house built for 8 or 10¢ ten years ago would run 12 to 18¢ now.

Dwellings, small frame, no plumbing ...... 5 to 6
    "   frame, good "   " .... 8 to 9
    "   " plumbing, heating, etc...10 to 12
    "   brick, plain, complete.........12 to 15
    "   " good   " ....15 to 20
Barns, small............................ 3 to 4
    " large (with stables)............... 4 to 5

Other approximate methods for preliminary estimates are the square foot method, based on the area in square feet covered by the plan, and the unit methods of cost per room, per animal to be housed, or per ton of material to be stored. Average costs per unit are listed below.

	Unit costs
Dwellings, per sq. ft.	$1.50 to $10.00
Stables, complete, per sq. ft.	$2.50 to $ 3.25
Dairy barns (large), per sq. ft.	$1.50 to $ 2.00
Residences, per room	$400.00
Stables (horse or dairy), per animal.	$200.00
Swine houses, per pen	$25.00
Hay storage barns, per ton capacity.	$12.00
Silos, per ton capacity	$1.50 to $4.50
Concrete floors and walks, per sq. ft.	7 to 12¢
Concrete work in place (no forms), per cu. yd	$5.00
Concrete work in place (forms), per cu. yd	$5.00 to $10.00
Concrete work in place (reinforced), per cu. yd	$8.00 to $10.00

Detailed Estimates.

The accurate method of estimating is to "take off" from the finished drawings, the amounts of all the different materials entering into the structure, and with ascertained local prices on materials and labor, add the totals of the cost of each item in place. To this total should be added say 10% for contingencies, and in the case of a general contractor, 10% for profits.

Materials are measured in the following units of measurement.

Lumber	Feet of board measure (B.M. = 1″ × 12″ × 12″)
Brick	Thousand (M)
Brick work	sq. ft. wall surface
Stone work	cu. ft. (or sometimes cord or perch)
Concrete and masonry	cu. yd.
Excavation	cu. yd.
Roofing	square (100 sq. ft.)
Paint	square
Shingles	thousand
Cement	bbl. (4 sacks)
Sand, gravel and crushed stone	yd. (cu. yd.) (sometimes sold by the ton)

As a general guide to present prices, the following figures, from actual costs, were used in the estimate of the dairy barn whose plans were given in Figs. 112 to 115. All figures refer to work in place.

Excavation	$0.30 per cu. yd.
Foundation	$7.20 per cu. yd.
Concrete floor	$0.11 per sq. ft.
Frame	$35.40 per M
Floor	$25.40 per M
Sheathing	$30.54 per M
Windows, average	$5.68 each
Doors	$0.48 per sq. ft.
Siding	$59.00 per M
Slate	$11.00 per sq.
Painting	$2.25 per sq.

The total cost of barn as estimated is $4320.00. In a cubic estimate this would give 5¢ per cubic foot, or $2.00 per square foot.

Heating, Lighting, Ventilation, Plumbing, and Sewage Disposal.

It is beyond the scope of this book to discuss these subjects, except to call attention to the need of considering them carefully in connection with the planning and design of farm structures, and to provide on the drawings for the systems proposed.

Thus in planning a residence we have to decide whether stoves and fireplaces will be used for heating, and if so to see that flues and chimneys are arranged in the most convenient and economical way; or if a hot-air or hot-water furnace is to be installed, to place furnace and fuel rooms to the best advantage for runs of pipe and location of air intake, and to locate registers or radiators where desired in each room.

In lighting, if gas or electric plants are to be installed, suitable provision must be made and shown on the plans for location and outlets. The advantage of an independent electric plant for both light and power for the entire farmstead should be considered, particularly if water power is obtainable.

The ventilation of dairy and horse barns has already been referred to, and is a very important consideration in modern construction. King's "Ventilation" should be procured and studied.

It is not usual in residence construction to provide a ventilating system, except as fresh air is introduced through furnace registers. The open fireplace has a distinct value as a ventilator, apart from its sentimental value as the spot about which home memories cluster.

A ventilating flue and hood is a desirable addition to the kitchen.

In planning for plumbing, pipes should be on inside walls if possible, and should for economy be kept in vertical runs. Thus the bath room is best located somewhere over the kitchen. There are various state codes to be observed regarding soil pipes, vents, etc.

In sewage disposal the advantage of the septic tank has already been referred. to. The dangerous cesspool should be abandoned. A general estimate of $30.00 per person for a family of ordinary size may be made in figuring the cost of an approved type.

Blue Printing.

Practically all working drawings are made in pencil on detail paper and traced on tracing cloth, a transparentized fabric, or sometimes on tracing paper. These tracings are blue-printed by exposing a piece of sensitized paper in contact with the tracing to sunlight or electric light in a printing frame made for the purpose. The resulting print has white lines on a blue ground. Any number of prints may be taken from one tracing. Blue print paper is bought ready sensitized and may be had in different weights and degrees of rapidity.

To make a blue print, lay the tracing in the frame with inked side toward the glass and place the unexposed paper on it with the sensitized surface against the tracing, Fig. 182. Lock up in the frame, seeing that no corners are turned under, and expose to sunlight for from one-half to three minutes, depending upon the speed of the paper. Then take out exposed paper and put it in a bath of water for about five minutes, and hang up to dry.

In almost any town one is able to have blue prints made at comparatively slight cost.

Problems.

The following problems are suggestive of a large number that can be made, using the data given, either alone or in connection with the illustrations or problems in Chapters III and IV.

PROBLEMS.

1. Hay is taken in at the end of a barn in loads varying from 500 to 1200 pounds. The track is supported by a yellow pine timber projecting 6'-0" without support. What should its dimensions be?

2. A grain bin with a capacity of 500 bushels of wheat is 12'-0" × 8'-0". The joists are of white pine spaced 12" from center to center. How large must they be to support the load safely?

3. How much cement, sand, and stone will be necessary to construct the water tank shown in Fig. 104 if its capacity is 15 barrels of water?

4. At $5.00 per cubic yard, estimate the cost of the foundation for the machine shed shown in Fig. 124. Footing 30" below grade. What will the floor cost at 9¢ per sq. ft.?

5. Silage (compacted) weighs approximately 40 pounds per cubic ft. How many tons of silage will be contained in a silo 14' × 38' if 6' is allowed for settlement after filling?

6. A farmer is feeding 40 steers the ration given on page 113. How large must his silo be to allow 200 days' feeding, supposing that the ration is the same for the whole period?

7. At 30¢ per cubic yard, what will be the cost of excavating for the house shown in Fig. 134? Make a sketch of wall section showing grade line, and basement 7' in the clear.

8. The dairy barn shown in Figs. 112 to 115 is to

be painted two coats. Allowing 1 gal. to 300 sq. ft. for priming coat and 1 gal. to 500 sq. ft. for second coat, how much will the paint cost at $2.00 per gallon?

9. Make a cubic estimate of the cost of the house shown in Fig. 134.

10. Make an estimate of the cost of the horse barn shown in Fig. 116. Use approximate methods to check your detailed estimate.

11. What will be the dimensions of a storage barn to hold 120 tons of hay? Make allowance for one-half of space above plate.

12. What is the weight of the wrought iron bolster iron shown in Fig. 168?

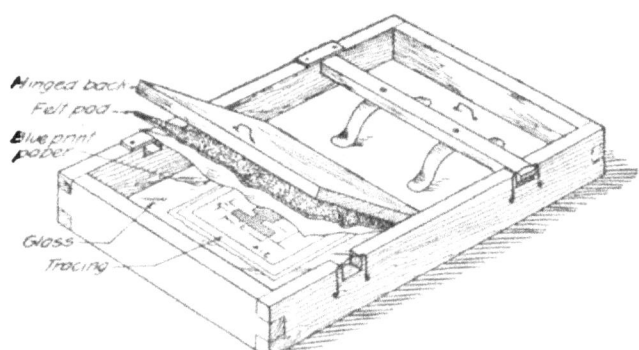

Fig. 182.—Blue-print frame.

CHAPTER VIII

SELECTED BIBLIOGRAPHY

A short list of books and pamphlets on allied subjects is here included, as a possible aid to those wishing to investigate further into some of the branches referred to in the present work.

It is not to be regarded at all as a complete list, but simply as a representative selection of recent publications.

On account of the rapid strides in scientific and sanitary research, and the use of new or improved building materials and methods of construction, a large portion of the books and other material now in print may be regarded as practically obsolete.

The list is classified into books; bulletins of the U. S. Department of Agriculture; bulletins of State Colleges and Agricultural Experiment Stations; and books or pamphlets issued by manufacturing concerns.

Government and state bulletins are published with some frequency, and much information may be gained from them. A monthly list of the bulletins published by the U. S. Department of Agriculture at Washington, D. C., may be had on application to the Editor and Chief, Division of Publications. State bulletins may be secured by applying to the Agricultural Experiment Station of the State.

Manufacturers are compiling valuable and interesting data and information in their catalogues of materials or appliances, and while intended for advertising purposes, some of these publications are well worth securing and preserving.

BOOKS.

Hopkins, Alfred.—Modern Farm Buildings.
206 pp. ill. $2.00. McBride Nast & Co., N. Y. 1913.

"Farm buildings of practical, sanitary and artistic lines." A book designed for large estates or country places, rather than the average farm.

Ekblaw, J. K. T.—Farm Structures.
347 pp. ill. $1.75. MacMillan & Co., N. Y. 1914.

A text book for Agricultural Engineering courses, with a general discussion of materials, construction and engineering of farm structures.

————.—Farm Buildings.
354 pp. 637 ill. $2.00. Sanders Publishing Co., Chicago. 1914.

A compilation of articles appearing in the Breeder's Gazette, contributed by practical men and dealing with actual structures.

Radford, W. A.—Practical Barn Plans.
172 pp. ill. $1.00. Radford Architectural Co., Chicago. 1909.

"A collection of common-sense plans of barns, out-buildings, and stock sheds."

————.—Poultry Houses and Fixtures.
95 pp. ill. $.50. Reliable Poultry Journal Pub. Co., Quincy, Ill. 1913.

A compilation of articles appearing in the Reliable Poultry Journal. "Poultry house suggestions for every breed, every climate, every fancy."

Child, Georgie Boynton.—The Efficient Kitchen.
242 pp. ill. $1.25. McBride Nast & Co. 1914.

A discussion of the planning of the kitchen from the standpoint of economy in operation.

————.—The Rochester Ice House.
Board of Health, Rochester, N. Y.

A bulletin describing in detail the construction and operation of the ice house illustrated in Fig. 127.

Ellis, A. R.—Making a Garage.
$.50. McBride Nast & Co. 1913.

One of the series of "House and Garden Making" books.

WHITE, C. E., JR.—Successful Houses and How to Build Them.

$2.00. MacMillan Co. 1912.

An extensive discussion of planning, specifications, fixtures and devices.

————.—Book of Little Houses.

107 pp. $.50. MacMillan & Co. 1914.

IVES, H. C.—Surveying Manual.

296 pp. $2.25. Jno. Wiley & Sons, N. Y. 1914.

A short treatise on surveying, especially designed for a brief course, or for home study.

ELLIOTT, C. G.—Practical Farm Drainage.

188 pp. $1.50. Jno. Wiley & Sons, N. Y. 1908.

The essentials, principles, and practice of drainage.

FORTIER, SAMUEL.—The Use of Water in Irrigation.

265 pp. $2.00. McGraw-Hill, N. Y. 1915.

Methods and structures pertaining to irrigation, principally of arid lands.

FRENCH, THOS. E.—A Manual of Engineering Drawing.

289 pp. $2.00. McGraw-Hill, N. Y. 1911.

A reference book on the different branches of technical drawing.

HANDBOOKS.

Valuable "pocket size" reference books with engineering data and information are published for architects, civil, mechanical and electrical engineers. The student who expects to work along these lines will find such a book almost necessary.

Bulletins U. S. Department of Agriculture.

BULLETIN No. 57.—Water Supply, Plumbing, and Sewage Disposal for Country Homes.

46 pp. 38 figs.

Discussion of Farm Sanitation.

FARMERS' BULLETIN No. 574.—Poultry House Construction.

20 pp. 13 figs.

Covers average conditions met in this field.

FARMERS' BULLETIN No. 589.—Homemade Silos.

47 pp. 37 figs.

Workable explanation of concrete, stave and modified Wisconsin types of Silo.

FARMERS' BULLETIN No. 438.—Hog Houses.

29 pp. 21 figs.

Tables of position of sun's rays at different times of the year.

Valuable for the design of sunlight type of swine house.

FARMERS' BULLETIN No. 607.—The Farm Kitchen as a Workshop.

20 pp. 6 figs.

Of interest principally to farm women.

FARMERS' BULLETIN No. 475.—Ice Houses.

20 pp. 11 figs.

Brief discussion of ice houses and application.

FARMERS' BULLETIN No. 623.—Ice Houses and the Use of Ice on the Dairy Farm.

24 pp. 17 figs.

The use of ice on the dairy farm for keeping of milk and cream in the best marketable condition is discussed. Suggested designs for six ice houses.

State Experiment Station and Agricultural College Bulletins.

BULLETIN No. 132.—Farm Poultry Houses. Agricultural Experiment Station Iowa State College, Ames, Iowa.

26 pp. 7 figs. 8 pl.

BULLETIN No. 152.—Movable Hog Houses. Iowa State College, Ames, Iowa.

43 pp. 37 figs.

BULLETIN No. 141.—Modern Silo Construction, Iowa State College, Ames, Iowa.

68 pp. 62 figs.

These three bulletins are excellent comprehensive treatments, in each case with working drawings and cost based on actual experience.

BULLETIN No. 179.—Construction and Equipment of Dairy Barn, Kentucky Agricultural Experiment Station, Lexington, Ky.

82 pp. 50 figs.

While this discussion is intended primarily for the State of Kentucky, it is wide enough to be of interest to dairymen generally.

MONTHLY BULLETIN No. 8, VOL. XII.— "The Missouri Silo." Missouri State Board of Agriculture, Columbia, Mo.

Well adapted to southern conditions.

Bulletin No. 110.—Georgia Experiment Station, Experiment, Ga.

26 pp. 9 figs.

Able discussion of types of silo suitable to region.

Bulletin No. 143. Economy of the Round Dairy Barn. Illinois Agricultural Experiment Station, Urbana, Ill.

44 pp. 42 figs.

Presents arguments for round barn as a type.

Bulletin No. 274.—Building Poultry Houses. Agricultural Experiment Station, Cornell University, Ithaca, N. Y.

44 pp. 66 figs.

Covers the subject in a thorough manner.

Trade Publications.

Lowden's Barn Plans. Lowden Mfg. Co., Fairfield, Iowa.

The James Way. (How to build a Dairy Barn.) James Mfg. Co., Fort Atkinson, Wis.

King, F. H.—Ventilation. King Ventilating Co., Owatonna, Minn.

126 pp. 63 figs.

A standard work on stable ventilation, adapted by a manufacturer of stable ventilators.

The Modern Farmer. Lehigh Portland Cement Co., Allentown, Pa.

64 pp. ill.

Small Farm Buildings of Concrete. Universal Portland Cement Co., Chicago, Ill.

158 pp. 136 figs.

Concrete in the Country. Alpha Portland Cement Co., Easton, Pa.

112 pp. ill.

Successful Stucco Houses. Clinton Wire Cloth Co., Clinton, Mass.

94 pp. ill.

Concrete Silos. Universal Portland Cement Co., Chicago, Ill.

104 pp. ill.

Concrete Construction, about the Home and on the Farm. Atlas Portland Cement Co., New York.

128 pp. ill.

The Concrete House and its Construction. American Association Portland Cement Mfrs.

$1.00.

INDEX